BLUE DRAGON CASE STUDIES

A Western guidebook to Eastern Medicine

Carol Ragle B.S., M.S., RN, D.O.M

Doctor of Oriental Medicine

This book is not intended to replace the services of a licensed health care provider in the diagnosis or treatment of illness or disease. Any application of the material set forth in the following pages is at the reader's discretion and sole responsibility.

ISBN 978-0-578-21967-7

DEDICATION:

"We are spiritual beings in a physical body-treat
that body like a precious possession"

"One Day at a Time"

"Take what you like, leave the rest"

"This is the uncommon path"

ACKNOWLEDGMENTS:

I wish to thank the "Mandate from Heaven" that pushed me to complete this work.

My school O.C.O.M Oregon College of Oriental Medicine, University of Northern New Mexico, University of New Mexico and Dakota Wesleyan University.

The blessed beings that made it possible, my editor Cindy, and housing to write...Loren, the ears to listen: Jerry, Linda, Carin, Susan, Paula, Mona and my loyal companions Regiman and MelleBelle........

ALL the clients whose stories I tell are the Heroes striving for Excellent Health!

With the power of your soul, anything is possible!

PREFACE:

This is a guidebook for Western oriented medical personnel, lay persons and general population. It is to help an individual navigate the Eastern thought processes around a western mind.

Western medical principles dictate a resolution, cure or fix around one particular issue. These medical principles have multiple layers of cause and effect. One cause fixed by one effect; example a drug or a surgery. Medical principle relies mainly on drugs, surgery for a fix on any given complaint (issue). When a person has multiple issues, stacking occurs in layers resulting in a bag of meds for multiple problems which can exacerbate each other.

Chinese Medicine takes a different approach of looking at multiple causes and fixing it with one treatment protocol. Chinese medicine breaks the body down into five organ sets; heart, liver, spleen (pancreas), lungs, and kidneys. There are two mystery sets of organs triple warmer (lymphatic) and pericardium (emotional). There are multiple ways of looking at these organs in combination which I will discuss later in the book.

Chinese Medicine has been around for 6,000 years. Many variations have been explored. Originally it was used to keep the Emperor alive so it is particularly useful for older persons. Western herbology approaches illness in the same way, one herb for one illness; Chinese herbology uses combinations of herbs to treat a trend in someone's health.

The following is a drug-free compendium to approach Western ailments from a Chinese view point. There will be repetition in various categories because of organ associations.

Chinese Medicine looks at a patient/client from an individual point of view, whereas western medicine has more of a cookbook approach to each ailment. What this means is Chinese medicine looks at the individual person vs. Western looks at the ailment.

Chinese Medicine relies on **evidence based** results from treatments which this book will show from 25 years of experience. Western relies on double blind studies that take years to complete.

Chinese Medicine doesn't need to know the cause to correct an issue, whereas Western medicine goes to great length to find a diagnosis so a cookbook cure can be used for it. With Chinese Medicine when there are multiple issues, the body sorts out the options and starts the healing process on the easiest to correct first. Western medicine doesn't work well with the gray areas; it tends to be black and white.

There are compendiums with hundreds of well researched herbs and herbal combinations in Chinese medicine like a PDR (Physician's Desk Reference).

Western medicine says it's only correct if written by a M.D.

Determination of a person's ill health or wellness:

Western: Lab testing, verbal interaction, physical therapy, and chiropractic.

Blood work = existing issue; The Chinese believe illness is a very serious issue when it is already showing up in the blood. The body tries to create homeostasis (balance) and will rob from organs to create homeostasis in the blood.

Chinese: The pulse, tongue, verbal interaction ("10,000 questions"), Shen (overall visual look). Prevention tactics before it shows up in the blood.

This is a practical guide A-Z reference for drug-free remedies using herbs, diet, nutrition, massage, homeopathy, acupressure and acupuncture.

The introduction explains the breakdown of oriental diagnosis vs. western medical diagnosis.

The following material describes common disorders, treatment and case studies.

There are forms on the website to determine organ imbalance and further information on particular problems.

Treatment Statistics were pulled from a clientele of 3,000 plus clients in New Mexico and South Dakota. There were differences between the needs between the two. In New Mexico D.O.M's are considered Primary Care Physicians and could send people for blood work and x rays. In South Dakota there still is no licensing for Acupuncture/Chinese Medicine, even though there are 47 states with licensure.

New Mexico showed up with more cancer, low back pain, seizures/epilepsy, spider bites.

South Dakota had more allergy, skin problems, autoimmune disorders and thyroid issues that showed up.

INTRODUCTION

The body is a highly sophisticated machine. I like to make the analogy that the body is like a battery in the car. One has to maintain the structure of that battery. Good food, nutrition, good habits and herbs keep the battery full and optimally working. Acupuncture charges that battery. Each body needs self-care and maintenance. We maintain our cars, house, and toys, everything deteriorates over time, Bodies included! Be proactive! Take "special care" especially as we age!

When someone comes to my office I evaluate the circumstances:

1 Why are you here?
2 What issues do you want fixed or resolved?
3 Are they acute or chronic issues?
4 Where is the breakdown in body parts?
5 Is it food sources or energy sources?
6 We work together to evaluate changes in diet, habits, emotional, spiritual or mental needs.

Chinese medicine asks for 12 treatments in general on an issue, usually 1 treatment a week. If extreme pain is an issue it can be treated 2 times a week up to everyday. I suggest 3 treatments and if it's not working for the client or myself we call it quits. If there is some good result, I suggest to keep doing it until resolution or sometimes one plateau's and then we know you can't go further with that issue. If there is no result after 3 treatments, 3 further treatments need to be done to see if it's chronic. If no result by then acupuncture probably isn't going to work. When people have multiple issues the easiest issue tends to resolve first. Sometimes people are good for a week and have the issue the day of treatment, this tells me it isn't resolved yet. When they go a week without issue, then I stretch it out for 2 weeks, when they are good there, they can go to 3 weeks. Some people just want to call for appointments if the issue comes up again. Some people like a maintenance schedule to come 1 time a week, 2 times a month or 1 time a month. Immune system issues are treated best every 3 weeks (21 days) an Immunoglobulin cycle. People come in according to their stress level, sometimes once a week, especially the Autoimmune or if one is working on a project like rehab or a compressed disc or low back pain, or tonifying a client to get pregnant.

The protocol I use is TCM: Traditional Chinese Medicine. We look at 5 organ pairs, 12 meridians, tongue, pulse and "10 questions" for initial evaluation.

Maintenance of body increases as one gets older. Maintaining our body functions, at optimal is imperative! Maintenance is preventative and Chinese Medicine can do that well to prevent early surgery, serious illness and just better quality of life.

Acute conditions are short term superficial and easily resolved. Chronic conditions can take years to resolve with many levels to address.

I use ITM (Institute for Tradition Medicine) for my Chinese herbal medications. I use Hannah Kroeger Homeopathic (in Boulder, Colorado) for viral problems. Standard Process is used for basic vitamins for thyroid issues. Enzyme International is used for liquid minerals. Nambudripad's Allergy Elimination Technique-NAET for allergy elimination and/or reduction.

The following cases are composites of clientele I have seen over the last 25 years, from an **evidence based** practice.

Schools of Chinese Medicine: There are four schools of Chinese Medicine reaching back 6,000 years.

First School: Tonify the stomach and spleen Qi. This means Chinese Medicine is used to balance the digestive system. This school believes you are digesting your food, separating the nutrients from the refuse and eliminating that which is not needed. As long as these three functions are working properly the body will perform at optimum capacity.

Dysfunctions Include: ulcers, acid reflux, esophageal construction, diarrhea, constipation, food allergies, diabetes, cholesterol issues, gout, headaches (HA's), migraines. Food allergies can effect arthritis, gout, diabetes, cholesterol, fibromyalgia, HA's, migraines, skin conditions, pregnancy and stress.

Second School: Tonify the Qi. This means Chinese Medicine boosts the immune system. This school believes if you have enough energy in your system (electricity) to run your organs all the organs will perform properly and maintain a properly functioning vehicle/body.

Dysfunctions Include: autoimmune, virus issues, herpes, mono, Hep C, fibromyalgia, allergies, asthma, chemical sensitivity, cancer, pain, M.S., lymph, mental illness, anxiety, arthritis, chronic fatigue, insomnia, hot/cold conditions, HBP, pregnancy, lung issues, stress and circulation.

Third School: Cleanse the system/body. This means eating good nutritious, wholesome food that doesn't toxify the body. This means eat what your body desires not what you emotionally want for a psychological crave. Usually cravings are for a mineral that the body lacks, like craving sweets is usually a need for magnesium. Eating protein when there is a crave for sweets usually fixes that issue. Eat proper foods for an organ that isn't working optimally. An example would be more greens for the liver if it's having cholesterol issues. The Chinese have the foods divided into groups from a color, taste and temperature perspective for certain organs. **The Rainbow Food Diet:** Colors are, Liver-green, Heart-red, Spleen-yellow, Kidney-Blue and Black, Lung-white. Doing cleanses also help rest certain organs so they can reboot; check Kroeger books for more information.

Dysfunction include: diverticulitis, crohn's, chronic constipation or diarrhea, cancer, lymph issues, cholesterol, diabetes, gout, gall stones, cystitis, parasites, thyroid, endocrine gland disorders, kidney disease, allergies, mental illness, arthritis, eyes, fibromyalgia, skin conditions, any form of toxicity, spider bites, virus and circulation.

Fourth School: Tonify the Kidneys. This means doing whatever it takes to improve the conditions of the kidneys. The Chinese believe that you are born with so much kidney Qi or energy, when you use that energy up, you die. The only way to replenish the Qi is to have good nutrition to build back up the kidney to optimal functioning. The problem is most people don't eat a very good diet and can't replenish, therefore their systems degrade until death. Most of my clientele come in at some point of kidney degradation, usually a variety of symptoms plus great fatigue. This is a constitutional issue. The overall system is degrading if there is no energy to keep it moving or sustained, like a car without gas.

Dysfunctions Include: All autoimmune disease, allergies, Alzheimer's, mental illness, anxiety, arthritis, any kind of pain, cancer, chemical sensitivity, chronic fatigue, digestion issues-"to cook the food", ears, fibromyalgia, incontinence, insomnia, seizures, epilepsy, skin conditions, vertigo, HBP, hot/cold issues, stroke, pregnancy, lungs/asthma, thyroid, endocrine glands, stress, circulation, teeth and bones.

All schools can work acute disease/superficial ailments; colds, cough, lung issues, stomach ache, strains, sprains, constipation diarrhea and UTI, and/or chronic conditions like diabetes, cholesterol, autoimmune, M.S, mono, virus, COPD, stroke, TIA's, incontinence and gallbladder.

Building the Kidney Qi back up is foundational. Chinese Medicine says that the kidneys in general help the ears in all aliments, the brain, grounds the heart, supports the pancreas (diabetes), strengthens the bones and teeth, supports hair on the head, rules ambition and will power, holds the center/builds essence (fertility) to conceive for woman and fertility for men. The **right kidney** gives day to day energy. It cooks the food in the digestive system. It sustains the lung energy to breath fully. It heats things up. It is in charge of the will to follow through on projects. The **left kidney** calms the liver therefore calming the heart (anxiety). It cools things down. It is in charge of the will to do things. Both kidneys (western) tell the body to absorb minerals (mineral steroids) into the bones and organs, to produce endorphins for the brain functions, to produce adrenalin/epinephrine (corticosteroids) to calm allergies, pain relief, regulates lymphatics, edema, heart hypertension, protein absorption, regulation of blood flow/pressure, blood purification, RBC formation, balance of acid/alkaline and urine production.

Contents:

Contents:

Contents:

ADD: Attention Deficit Disorder

ADD comes into my office between the ages of kindergarten to 14 years old; more often with allergies and usually with the parents who don't like how the prescription drugs are affecting their children. The kids can't sleep or eat and have flat emotions. **Treatment:** NAET, diet and homeopathic.

Analysis: Correcting the absorption of nutrients into the child's body can correct a lot of symptoms. The dietary change to foods that don't create heat; no sugar or caffeine. Use kidney foods to support a depleted adrenal system like meat, root vegetables and blue foods. Quiet time with soft music or meditation helps calm them and get them into a regular practice of self-care. Learning to self-monitor is helpful and a life-long endeavor. When older, acupuncture can help balance out the energy in the body for more wholesome function.

Note: ADHD included in its own category.

Addictions

Alcohol, drugs legal and illegal, and food addiction can all be helped with Chinese Medicine. There is a standing protocol and organization developed just for addictions. The organization is called NADA-National Acupuncture Detox Association. They have had great success with their programs across the nation. They have used the program for State Psychiatric wards, drug court, meth addicted pregnant women. At one time there was a hospital in NYC dedicated only to addictions of all kinds. Five needles are placed in the ears at certain points, lungs, liver, kidney, Shen Men-for calming, sympathetic-for calming the nervous systems. I worked in school at a drug court center for a year doing counseling and education for living in the world. It was a great program and many people (80%) were able to achieve independence from their addictions and able to thrive in the real world.

ADHD: Attention Deficit Hyperactive Disorder

ADHD Case Study: A 31 year old female presents with ADHD, anxiety, allergies, insomnia, migraines, emotional, frustrated, sad, happy and anxious. **Treatment:** Acupuncture one time a week until more settled. She took ITM Restorative to tonify the kidneys, ITM Acorus for anxiety, homeopathic for virus and spider bites.

Analysis: After three acupuncture treatments the anxiety was highly reduced, she was feeling well. Sleep was better with acupuncture relaxing the muscles and anxious brain. Detox off of two viruses was getting the pain out of the body. When a sensitive body is fired up long term they usually get the diagnosis of ADHD because the brain cannot settle the body down, the body is hyped up because something is wrong. Deep breathing, meditation, acupuncture, herbs, essential oils and even stones can help ground and center an ADD/ADHD person. They usually present with a lot of allergies that keep the body agitated contributing to migraines and emotional frustrations. She additionally had miasms of being bit by a brown recluse spider three times. Spider venom creates constrictions in the body, tight muscles that can't relax, migraine headaches, emotional anxiety and digestive issues.

Allergies

ADHD Case Study: A 9 year old female presents with ADHD, Hyperkinetic, belching, gas, sleeps a lot, allergies, ear tubes-twice, tonsil and adenoid surgery. She was on medication for one year and it didn't help and was totally exhausted with the pill. **Treatment:** NAET for allergies, DHEA for adrenal fatigue and kava herbs for sleep.

Analysis: She presented with food and pollen allergies which hyped up her body so she couldn't focus and sit still. Vitamin C was a trigger that kept her up all night. Calcium mix gave her a headache. She craved sugar. After caffeine, grain, yeast, formaldehyde, plastic, minerals, fats, artificial color and additives were treated, she could sit still and do her homework and go to church quietly. She got into less trouble at school. The girl was pleased and so was her mother, five years later, still happily having a good school year.

Universal Reactor Case Study: A 78 year old female presented with allergies to everything a "Universal Reactor". She had fibromyalgia, dry eyes, cold body, shortness of breath, low energy, insomnia, white diet, exposed to pesticide poisoning and took her ten years to detox. She takes a lot of natural remedies to keep her balanced. **Treatment:** Acupuncture, NAET and homeopathic.

Analysis: We had to go really slow through the NAET process because her kidneys would detox quickly and throw her into an adrenal detox reaction that would make her sick until it cleared; then she felt much better and could absorb her food. She had viruses that needed to be removed slowly one per every 1-2 months so she could recover her immune system.

Lips Case Study: A 59 year old female presented with lips swelling over six weeks prior to coming to me. She used hydrocortisone steroids, mupirocin, bacitracin and Vaseline to try to calm the lips down. She had shortness of breath with climbing stairs, heartburn, constipation and occasional hot sweats even though she was through menopause for six years. In the past she had migraines with chocolate. She can wait five days before moving bowels, tends to be hot at night. Avocados make her ears feel funny. **Treatment:** ITM

Restorative to control the heat to calm the system, helps with heartburn/heat and cools the bowels. She flared with strawberries; we tested her for allergies using the NAET system. She improved dramatically after Vitamin C and even better after calcium mix. We also cleared her on grain mix, yeast mix and finally salt mix. She had no more apparent problems with her lips, even though, she had more allergies that she did not clear with NAET.

Analysis: Restorative herb cools the liver to calm the heartburn. When the bowels are too hot they get dried out and create constipation. Chinese diagnosis always determines if someone is too hot or too cold. The body constitutional temperature can push the immune system, allergies, and digestive system into dysfunction. The NAET reprograms the brain so a person isn't allergic to a particular allergen. She came in for six treatments, two acupuncture for the body and four NAETs.

Pollen Case Study: A 14 year old boy presented with red eyes-blood shot, tears running down his face, sneezing, blowing nose, scratching his eyes, hot, frontal headaches, no energy, belching, sleeps nine hours a night, can be irritable and depressed. Allergies got worse after an extra dose of Hep B and Gardasil. He was on Claritin and VyVanse for emotional disorder. **Treatment:** NAET

Analysis: By the time he finished the first treatment his eyes had quit watering and stopped itching. We used weed mix (150 weeds). We proceeded to do the rest of the pollens, which worked so well he wanted to do the foods. Five years later he was in college getting his degree in accounting; he could focus. He was able to absorb the nutrition to fuel his brain so he was able to function normally. Allergy disrupts the brain from functioning properly if one isn't absorbing the nutrients that build the brain endorphins.

Peanuts Case Study: An 8 year old boy presented with a peanut allergy since two years old. Issues with cat, milk, eggs, cheese, shellfish-swollen tongue, tends to be hot, high energy, eats lots of milk products, sweets and pasta. **Treatment:** NAET

Analysis: He was treated for his food allergies, treating the less active ones before we did peanuts. Peanuts are grown underground creating a possible mold allergy instead of a peanut allergy. His high consumption of sweets and pasta can contribute to the heat in his body which can contribute to high activity or in some people agitation. We treated all allergens and lastly peanuts. He decided he didn't like peanuts after all, but could eat them now, which protected him from anaphylaxis. He was able to move from the allergy table to be with other friends socially. He thrived to become quite popular.

Foreign Born Case Study: A 3 ½ year old boy presented with food issues coming from an orphanage where mono meals were presented with no variety. The parents were concerned. He was not interested in vegetables. No history of vaccines. Diagnosis was food allergies, parasites, difficult birth, and emotional trauma. He developed ringworm which peeled after treatment of grain and yeast mix. **Treatment**: NAET and herbs: ITM Omphalia for parasites, and TB residue for a cough issue.

Analysis: One can be truly allergic found in a blood test or intolerant or malabsorption found by kinesiology. His intolerances were essential to identify so he could eat what would absorb and grow him to adulthood. He was intolerant to soybeans, nightshade-potatoes, tomatoes, peppers, sugar, and cow's milk. He was okay with goat milk, onion, garlic, and eggplant. Parasite cleanse was essential to optimize his growth. He had at least four kinds of parasites. Omphalia is an herbal taken two times a day for at least three weeks to get the parasite parent and the offspring. Use for four to six weeks for optimal results.

Mononucleosis Case Study: A 9 year old female presented with strep throat after mono, dark circles under eyes, bumps on skin, dizzy when hot, urinates a lot, thirsty, all vaccines were given. She was on probiotics. **Treatment:** NAET, homeopathic: Mono/Epstein Barr.

Analysis: Mono is a gateway virus that opens the body up to multiple issues. One can develop allergies or more viruses jump on to create fibromyalgia, autoimmune, chronic fatigue, diabetes, nerve pain, thyroid, herpes, shingles, molluscum. mono can jump on a weak immune system from allergies. A person can have mono life long and develop into cancer from lack of nutrition and too many untreated allergies. She got through all the NAETS and the homeopathic for mono within four months, at six months she was happy, healthy, and active.

Anemia

Anemia is a red blood cell deficiency. The red blood cells take the oxygen around the body to invigorate the tissues so they work well. A shortage of red blood cells can make a person short of breath, lethargic, loss of appetite, appear pale, brain fog, trauma, bruising or bleeding from an accident can cause blood loss. Unseen issues can be that the food is not metabolized to get the nutrients to make red blood cells or the kidney is so weak it doesn't tell bones to make red blood cells. Low thyroid can aggravate this condition.

I had five primary complaints of anemia, average age was 54 years old, all complained of fatigue, brain fog and weakness. One did not respond to iron so we did NAET. One had a thyroidectomy making regulation of the thyroid through medication more difficult. There were two men and three women. **Treatment:** All did at least ten acupuncture treatments and some ITM herbs.

Analysis: Virus needs to be checked in all cases to see if the viruses are depressing the immune system. NAET is helpful to get the base nutrients available for absorption.

Acupuncture charges the immune system and kidneys so blood can be produced. It takes between 12-16 weeks to replenish the blood.

Anxiety

Anxiety Work Related Case Study: A 58 year old female presents with shoulder pain that radiates to her fourth and fifth fingers, mid-back pain, heartburn, anxiety, stressful work, frontal headaches, sinus, and insomnia. She had a full hysterectomy because of a prolapse. Her diet was reasonable eating every type color of food. She has bouts of depression and sadness after surgery. Her heart races with anxiety. She has eye dryness, constipation, and with heartburn belching. **Treatment:** Acupuncture and ITM Restorative herbs Kroeger homeopathy.

Analysis: Restorative herbs were used to clear the heat in her system. She was in a very responsible position of managing others in a very fast paced environment/critical care. Anxiety in Chinese Medicine relates to the liver and the heart. When the kidneys can't calm the liver enough to calm the heart anxiety occurs. The heat in her system creates the heartburn in her digestion, insomnia, eye dryness, and constipation. She was recommended olive oil for the dry eyes to put on her skin instead of lotion after showers to get more good oil in her system. Anxiety can burn up the oils in the body (endorphins) because of the stress. These people need more good oil in their systems. Oil keeps the moisture in the skin and eyes. She ate a rainbow of foods, which Chinese Medicine suggests; one eats every color every day. The pain in shoulder radiating to the fingers was alleviated by acupuncture. Her headaches improved. Sinus and heartburn were also addressed with NAET. No complaint with heart racing after two months. Bouts of sadness and depression alleviated when on vacation. She did acupuncture for anxiety every two weeks; then every month for six years until she retired.

Adrenal Fatigue Case Study: A 21 year old male presents with insomnia with three hours of sleep while training with work, testosterone up, cortisol up, has a cold for three to four weeks, eyes dilated, exhausted-can't turn off, digestion is okay, exercises with stress. **Treatment:** Acupuncture with ITM Restorative herbs.

Analysis: This client has so much heat in his system he can't sleep and has moved from adrenal fatigue to adrenal exhaustion. After two treatments he settled down enough to sleep and became more emotionally balanced with a better mood. He used exercise to destress, however if one over exercises it can run the kidneys down from fatigue to exhaustion. Dilated eyes indicate adrenal exhaustion. He would come in with dilated eyes then go out with normal pin points. I asked him to watch his eyes and when dilated he needed to rest. Kidney energy improves with rest. The Chinese believe one should rest as much as one is active. He did a total of four treatments and corrected his diet so he wasn't amped up by it. (excess sugar, caffeine and certain amino acid supplements.)

Arthritis

Rheumatoid Case Study: A 60 year old female presented with crippling Rheumatoid arthritis since 24 years old, survived breast cancer twice, she was hot, heart palpitations, and ringing ears. She exercised and ate a reasonable diet. She was on Cozaar, Toprol, Evista, garlic, fish oil, Multi-vitamin, Vitamin C, Vitamin E, Folic acid, Magnesium, Benadryl, and took aspirin for pain control. **Treatment:** Acupuncture and ITM herbs, Kroegar homeopathic and NAET.

Analysis: We boosted her immune system for three months with acupuncture, and then proceeded with NAET. In Chinese Medicine arthritis is usually a food allergy issue. There is either intolerance or malabsorption. We even had to treat her medications for arthritis and heart palpitations. She had viruses, parasites, and toxicity...she was in a TANGLE. (see index) We slowly detoxed her from the parasites first, the viruses one at a time and kept her immune system supported throughout the treatments with acupuncture. Cancer devastates a body and leaves it vulnerable to the tangles. Cancer treatment can make a person more intolerant and less absorptive of foods. Care needs to be taken when reconstituting the body that has gone through so many traumas. She needed a lot of immune support and good diet, which this woman was doing. It took three years to get this client functioning better with less pain. She had more energy and was active seven years later.

Osteoporosis Case Study: A 56 year old female presented with hip pain that radiates to her ankle, neuropathy in feet, digestive issues, hot, low energy, knee pain, sweaty, migraines with menses, insomnia, total hysterectomy due to fibroids. **Treatment:** Acupuncture, ITM herbs Restorative, Women's Treasure, Lysmachia, Liquid Minerals and Echinacea.

Analysis: Her malabsorption of nutrients does not allow for absorption of raw material to correct the pain that is created by Osteo arthritis. We boosted her system; more energy so she could cook her food better. The acupuncture stimulates release of natural endorphins to relieve the pain. It also relaxes the muscles around the hip to release the compressed nerves that create neuropathy. Restorative herbs corrected the hot, sweaty, insomnia issues. The knees improve as the kidney energy increased. The liquid minerals satisfied the bodies crave for more nutrition in the bones that ached, for example Osteoporosis Arthritis. Echinacea was used for cold or flu, it stops viral replication. Lysmachia was used for kidney stones that was causing her back ache. The herbs soften up any stones in the body so they can be excreted. Women's Treasure herbs were used to increase her day to day energy. It took four months to get most of the issues resolved and then she went to maintenance of every 2-3 weeks. She was under a lot of stress with her family owned business and Acupuncture helped tremendously.

Asthma

Asthma Case Study: A 24 year old female presents with asthma, allergies whole life, wheezing, digestion is slow, irregular periods and long time frames between, hair on head grows slow. She took allergy shots in past and is allergic to animals, pollen, mold, dust, and some food. Her medications include Advair, albuterol PRN, probiotics, calcium, and prenatal vitamins. **Treatment:** Acupuncture, ITM Women's Treasure, Nuphar Perilla seed, Xanthum, and NAET.

Analysis: Asthma in Chinese Medicine relates to the Kidney Qi not grabbing the lung energy therefore shortness of breath. In this case there is a lifetime of allergies that drain the kidney, overuse of natural steroids to neutralize the allergy reactions. There are not enough natural steroids for lung function so the client has to do medications that support the lung/kidney interaction to breath. Right kidney helps cook the food and regulates the menses. She has slow digestion and irregular, long between menses cycles. Hair on the head is slow to grow, another kidney function. She needed kidney support first with acupuncture and ITM Women's Treasure. Perilla seed was for the lungs to ventilate the wheezing. Nuphar was given later to regulate the menses. Xanthum given for sinus and allergy issues. After she was stable we started NAET for her allergies. She was able to get off all her medications with an occasional allergy break through; she would use her meds then. Stress aggravated her asthma and also congestive colds. Over a year she did NAETs and would return if necessary for energy boosts. New allergies occasionally show up, however she hasn't been back to the office for five years.

Autism

Autism: It is controversial in the medical profession what causes autism. I have treated 15 individuals with varying levels of autism and varying ages; 14 males and 1 female. Alcoholic and drug addicted mothers can produce autism and other childhood maladies, for example downs syndrome. The majority of the children I treated had a reaction to a vaccine. Every parent claimed the child changed within days after the MMR vaccine. (See Immune System and Vaccines). Acupuncture was used in older individuals and can help stabilize the body. NAET was used to reprogram allergy issues. NAET produced the best results in that the children started functioning at a higher cognitive level returning to normal childhood development.

Autism Case Study: A client presents with a very active, hot, 4 year old child. He paces, repeats words and sings a lot. Parents saw a change in behavior after a reaction to MMR vaccine. He has a poor appetite and craves sugar, salt and pickles. **Treatment:** NAET

 Analysis: After the first treatment of brain/balance/body with calcium mix he could sit down longer every day and played more with other kids. He had more energy and progressively started speaking more after three NAET's of vitamin C, grain, yeast mix and minerals. After minerals he could say "daddy, daddy" and could speak clearer words. After six NAET treatments he was eating more and qualified at 50% growth, mother was pleased with progress. After more NAET's he gave more eye contact, was asking for the potty, playing with kids, connecting thoughts and communicating them. After 17 NAET's over 6 months he was focusing and doing better in school, he was trying to do more.

Auto Immune Disease

Auto Immune: 14 treated, 11 females, 3 males, average age 50 years old

Auto Immune is complex. To work out of it one needs to stage the levels of illness. While building the depressed immune system with acupuncture, herbs, good diet, and good habits one can remove obstacles of viruses, yeast/candida, parasites and toxins in the body. One has to be careful with cleansing gently and very slow detox. NAET helps rebalance malabsorption and homeopathy can pull out the virus and toxins over 1 month each, slowly; so a person can function as they improve their health. It is very easy to crash an autoimmune person. The process needs to be slow and take great care. A person can have several autoimmune diseases at a time.

Autoimmune includes Raynauds-circulation (hands and feet), Grave's Disease/Hashimoto's (thyroid), Chronic Fatigue (Kidneys), M.S.-muscle fatigue (Liver), Lupus-excess heat in the body, hyped up immune system, Parkinson's-toxicity of organs and tissues, Scleroderma-skin issues/circulation.

Hashimoto's and Grave's Disease: always has an underlying kidney component, adrenal fatigue. The thyroid and adrenals have an antagonistic relationship. The thyroid wants to hype up the metabolism. The kidney wants to soothe, calm the system.

Lupus: is common in autoimmune with chronic fatigue. Rosacea is the common symptom, rosy cheeks that don't diminish in color with bumps, pimples, irritations. This is easy to fix in Chinese Medicine. This syndrome is an excess heat issue in the body. Look for the constitutional issue that is creating it. 4 treated all women at an average age of 47 years old.

Parkinson's: is usually a chemical toxicity of some sort. 7 treated 3 females, 4 males, average age of 69 years old. Originally alternative medicine thought it was associated with

Alzheimer's disease and overabundance of aluminum in deodorants and other toxic chemical exposure. It usually occurs in an already compromised situation with some other trauma, surgery, leukemia, valley fever, diabetes, and osteoporosis.

Chronic Fatigue: Treated 42 people, 35 females, 7 males, average age 45 years old. Chronic Fatigue is a collection of symptoms predominating in adrenal fatigue/exhaustion. Many diseases have this component in it which I discussed in "Kidney School" throughout this book.

Fibromyalgia: can be included in Chronic Fatigue Syndrome category. Usually there is underlying mono that opens the gate to other viruses. Poor diet, bad habits, crazy lifestyle can burn out a body. Over work, too much exercise, over vaccinated, too much sex, over sensitivity to life, over emotional all drain the kidney/adrenal complex.

M.S.: 15 treated, all female, average age of 47 years old. In Chinese Medicine M.S. relates to the lack of innervation to the muscular system which relates to the liver. I have found that there is a viral overload that initiates the cascade. I had 3 nurses all started the Hep B shots, took 3 in a row when they first came out and within 3-6 months each of them came in with a diagnosis of M.S. Acupuncture and herbs can help maintain the immune system at a fairly high functioning rate but I find that this disease is not removable even with homeopathy or NAET's. As long as a maintenance schedule is maintained there is usually sustainability. Other autoimmune can be brought up to stand alone without continued maintenance. Maintenance is always recommended for anyone who has ever had any autoimmune problems.

The underlying allergies can be helped by acupuncture and herbs to boost the immune system.

Chronic Fatigue "Sick Building" Case Study:
A 36 year old female presents with overwhelming fatigue, could only do one thing per day and go back to bed, hot,

sleeps a lot, chemically sensitive to many smells, formaldehyde, molds, perfumes, food sensitivity to many foods, milk products, grain, yeast, pollens, SAD (seasonal affective disorder). Reacted to a "sick building" high rise, little fresh air return (5%) and mold two feet up the walls in the basement. New carpet put in and took her down. Emotionally over reactive, depressed, sad and angry lost her job and in litigation for work compensation. She has seen many doctors, some understood, some feel she is slacking. **Treatment:** Acupuncture, ITM herbs, Kroeger homeopathic, NAET.

Analysis: "Sick Building Syndrome" is controversial in the medical circles. This building was in Alaska and the culminating incident occurred in the winter when there was only 5% fresh air return in a 3 story building. Water damage had occurred and there was 2 feet of mold on the walls in the basement with 2" of standing water. A chiropractor's office had "open air" Quadraldehyde containers open to air and filtered through the central air system. New carpet sprayed with formaldehyde and glued to the floor added more chemicals to the air. It was a chemical soup creating chemical sensitivity and food sensitivity. Chemical sensitivity can create heat in the body because of the hyperactive immune system leading to insomnia and fatigue due to the wearing out of the kidneys (adrenals) immune response. The kidneys tell the bones to make the blood, red and white blood cells. Long term the body gets very fatigued with kidney overuse; kidneys rule the energy. Acupuncture for four years plus herbs got the body back. Homeopathy got the viruses out, six of them, including mono. NAET to reprogram body not be allergic to chemicals, pollens, molds and foods. This is my story and I lived to tell.

Graves Case Study: A 35 year old female presented with Graves's Disease for 15 years. She has allergies, fibromyalgia, heart palpitations, low energy, tends to be hot then cold, gets sinus headaches with dust, ears irritated with cold, has anxiety, eats a balanced diet, tries to eat gluten free, no alcohol or caffeine, on birth control now because 3 years with irregular menses. She works on a farm/ranch. She is taking Propranolol, Synthroid, Iodine, Selenium, D3, Calcium, Magnesium, Zinc and Multi Vitamin. **Treatment:** Acupuncture, ITM herbs, NAET

Analysis: ITM herb, Restorative for energy and to regulate hot/cold symptoms. 17 NAET's to correct underlying allergy issues for Grave's Disease. After 3 NAET (Vitamin C, Cal mix, Animal Fat) treatments, sinus problems were better, after minerals she had no headaches or achy joints, energy was good after 7 treatments and is now off sinus medications. She can breathe better after all pollens and mold were treated.

M.S Case Study: A 44 year old female presented with M.S. Spasticity of hands and legs, lack of bladder control, dizzy with fatigue and low energy. She has had numbness in hands for 11 years and now spasticity moving into her hands, last year it was into her legs. She is worse with fatigue and better in the morning, worse with heat, spasticity in the eyes. 18 years ago diagnosed with carpal tunnel. Disease started with stress from the death of her dog. Her body is hot and hands and feet warm. Menopausal, no problem with menses, no children, tried chemotherapy 2 times and got worse spasticity. Eats out a lot, lots of fried food, likes sweets, does weights for upper body strength. Takes Baclofen for spasticity, Amantadine for energy, Multi Vitamin and B12. **Treatment:** Acupuncture, ITM herbs Restorative and Gecko, homeopathy

Analysis: Herbs to charge up kidneys for energy. After 4 acupuncture treatments she could write better, right foot could lift up, energy better for a day, bladder got better at eliminating/stronger stream. She started to do more and balance was better. She started to watch her foods. At 10 treatments 1 time a week she was able to rollover in bed- new accomplishment; since original onset of disease. Started using tens machine to get more out of her leg and arm. She tried at 6 months to stretch treatments out to every 2 weeks and it didn't work. She improved slowly over the next 6 months; she still needed 1 treatment a week. We did some parasite removal, virus detox. We used ITM herb Restorative to clear the heat in the hot body, hands and feet. Her quality of life was improved. I moved from the area. I had heard she died a year later.

Back Pain

Back Pain: 331 treated 188 females, 144 males, average age of 55 years old.

Back pain is my largest category of patients. They come in for scoliosis, sciatic (see low back pain), compressed discs, bulging discs, stenosis, gallbladder/back pain, after care for surgery repair, lumbar and cervical pain, spasms, tailbone compression, hip replacement, ruptured discs. Acupuncture works the best at relieving muscle contraction around the area of pain. Distal areas are needled to pull the pain away from the offending area. Low back relates to kidneys, so kidney support needs to be addressed. The Chinese believe if an organ is weak, then the area that represents that organ will be weak. The spine has a vertebra that represents each organ.

Chronic Pain Case Study: A 70 year old presents with chronic pain "all bones broken" ex-sheriff and ranch hand hard worker, abuse on the body due to professions plus accidents over the life time. Right broken foot, spinal fusions, shoulder and hand pain. He was put on Fentanyl patches, Tramadol, Cymbalta, Oxycodone, Advil for pain, stool softeners for the constipation that is created by the opioids. This man could hardly walk in the door, shuffling to the table; hard for him to get on and off the table. He had a hard time going through withdrawals before Chinese Medicine and would get spacey, uncontrollable shakes and pain. It took three months to get off Tramadol and did so without withdrawal or signs and symptoms. At that time he was able to get on and off the table easily without assistance. His mood improved and less pain overall. Fentanyl is a harder issue. He has a new patch every three days and is trying to stretch that out further. He feels better and is doing more. He had acupuncture every week for one year at time of writing; he had tried Turmeric, Curcumin, and CBD oil with some success for pain. **Treatment:** Acupuncture and ITM herbs

Analysis: Acupuncture stimulates the endorphin receptor sites so a drug is not needed to complete the chemistry. Acupuncture helps the body to produce its own endorphins to block the receptor sites. The body will create tolerance to opioids so one needs to take more and more to neutralize the pain, eventually one dies from the overdose or has to detox to start the cycle again. Detox can kill a person because of dehydration and severe withdrawal reactions such as vomiting, diarrhea, and general malaise. Pain is the body's way of telling a person something is wrong. Long term pain control masks other problems in the body that need to be addressed. Acupuncture can help the other parts of the body to strength while helping the body to detox.

Scoliosis Case Study: A 70 year old female presented with scoliosis her whole life, right hip pain at night sometimes radiates down leg, diagnosed with arthritis, hashimoto's disease, shortness of breath up and down stairs, heart palpations, likes sweets, milk products, diet soda, had hysterectomy for an ovarian tumor. Medications include: simvastatin for cholesterol, meloxicam for arthritis, levothyroxine for thyroid, and motrin for pain.

Analysis: Initially she felt better three days and then the pain was back the next week the treatment lasted the full week. She became less tired and the right hip had resolved after six acupuncture treatments.

Note: Scoliosis is a chronic long term affair. Most people do maintenance with acupuncture over time to keep the muscles loosened up or they do massage.

Low Back Pain Case Study: An 83 year old female presents with low back pain which radiates down both legs to the knees. Her eyes are blurry and red; states she "can't see" but did have cataract surgery. She's light headed, constipated, sweats, has restless legs, her bunions hurt, she is hot with cold hands and feet. She craves sweets and eats them. She eats high protein with fruit. She has sleep apnea and uses oxygen at night. She is on hydrocodone, gabapentin, citalopram, alprazolam, calcium and eye vitamins. **Treatment:** Acupuncture and ITM herbs.

Analysis: After five treatments she was able to drop the hydrocodone for pain, her balance was better, bowels improved, restless legs improved with liquid trace minerals, energy increased eyes still bothered, bunions improved after two treatments, no more pain. A year later her back was fairly good, toes on and off but the eyes were still bad. She was on more medication five years later adding klonazepam and metformin. A lot of these drugs create heat in the liver which rules the eyes. Her red, blurry, irritated eyes went into macular degeneration. She now had polymyalgia, which is in the fibromyalgia family. She had muscle pain all over. Muscles are also ruled by the liver. She said she was achy from head to toe. She was on so many heat producing, liver aggravating medicines for pain that she was beyond tolerance for the liver. Her eyes were being damaged and she ached all over. I suggested she detox some of the medications to see if she could recover some of her eye sight and get pain relief.

Sciatic Case Study: A 70 year old female presented with right hip sciatic, low back pain, no energy, cold, frontal headaches, occasional eye redness, hard to get to sleep, drinks diet pepsi (3)-16 onces daily, artificial diet-ice cream daily, and cholesterol issues. Medications include: simvastatin for five years, Zocor, libandronate for bones, multivitamins, biotin, tylenol and advil for pain. **Treatment:** Acupuncture and ITM Women's Treasure.

Analysis: After three acupuncture treatments the pain went away. She stopped the statins after the first treatment and reduced the diet pepsi, artificial sweeteners and caffeine that caused heat in the liver. A hot liver can cause eye redness. A person can generate more cholesterol in the liver to protect the liver. Sweeteners aggravate the liver because the liver doesn't know what to do with the chemical because it's not natural. Artificial sweetener aspartamine is a known carcinogen. Statin side effects are known to cause muscle pain especially in the legs. One can go off them and see if they are the cause without withdrawal. Any clients with long term use of statins who present with leg pain are suspect. Women's treasure herbs help with energy, body coldness and frontal headaches.

Low Back Pain Case Study: An 87 year old male presents with low back pain, loss of balance, dizzy, insomnia, walking is a problem on the ranch, eats a lot of sweets, stress financially. Medications include: omega 3, CoQ10, selenium, calcium, magnesium, zinc, vitamin D, nitro dur, metoprolol. **Treatment:** Acupuncture and ITM herbs Gecko.

Analysis: After four treatments he went off metoprolol and became less dizzy. Sometimes the elderly become more sensitized to their medications. The medication becomes too effective, in this case brought his blood pressure too low therefore off balance and dizzy. His balance started getting better and better. The gecko herbs were used for his energy. After seven treatments he walked a straight line. After four months his back was good, fair energy level and better balance.

Bulging Disc Case Study: A 45 year old male presented with grabbing back pain, x rays diagnosed arthritis with compressed disc, right side pain/pinching. His body temperature is hot, has allergies to pollens and molds seasonally, has headaches at the base of the head with agitated sleep due to work, occasional heart palpitations. Diet consists of meat, veggies, little fruit, a lot of milk and cheese, he likes sweets. **Treatment:** Acupuncture, ITM herbs

Analysis: He was put on ITM herb Restorative to clear the heat that contributed to headaches (heat rising from the back pain to the head). The herbs also helped calm the heart palpitations and lack of sleep due to heat/pain. Acupuncture used for 6 treatments to relax the muscles at the bulging disc, clear the heat, calm the spirit, and reduce pain. 5 years later I saw him at the store and he said he felt great! That was his last chance before surgery, he was a contractor.

Bell's Palsy

Bell's palsy is treated with acupuncture on the face. Bell's palsy occurs on one side of the face or other. The face droops with lack of nerve stimulation. The nerve is called the trigeminal. This nerve reaches up over the eye to the temple, branches out to the eye, cheek, and lower jaw, therefore the eye droops and the mouth goes flaccid. These three branches start at the top of where the ear attaches to the head and becomes the primary spot to needle.

This phenomenon can occur with great stress, too much heat/pain in the head, neck, shoulders, and from viruses. Suppressed emotions can irritate the issue.

Analysis: Out of eight patients, six patients resolved the droopy face with an average of ten treatments. Three of these had benign tumors on the back of the ear surrounding the nerve and after surgery developed the issue. They were all female.

Bone Spurs

Baker's Cyst: This is a fatty deposit of protein that the body creates to protect itself behind the knees due to some trauma or stress on the knee. It is benign, meaning non-cancerous.

Fatty Tumor: This can be created anywhere in the body, the feet, hands, back, and head. They are usually created to protect an area from trauma, like a nerve from a rough edged bone. **Bone spurs** can also be created by the body to help support the bone structure if there is weakness in an area, for example heel spurs. **Treatment:** The fatty tumors and cysts are treated by a technique of "surround the dragon". A needle is placed at the four corners of the problem and one in the middle. I have seen the tumors go away completely after a month of one treatment a week and some are greatly reduced with regular care but never completely disappear. Removal of cause is usually the cure for the ones that don't go away, which includes further investigation with an M.D, x rays, etc.

Herbs work well for ovarian cysts, plus acupuncture. Bone spurs take about six weeks of treatment and rest which, most people don't have time for so they opt for surgery.

Bowels

Constipation Case Study: A 2 year old boy presents with constipation since he was born, colicky, back pain, leg pain, lactose intolerant, trouble with many formulas, no celiac, sick often with strep virus, irritable emotionally, a lot of sugar products in diet, taking probiotics and lactulose. Allergic to almonds, breaks out on the face. **Treatment:** NAET

Analysis: Treated with NAET Calcium mix, he was constipated for three days, after that regular. After vitamin C, two hard bowel movements, then regular. He was treated for grains, nuts and the constipation resolved.

Ulcerative Colitis Case Study: A 58 year old male presents with Ulcerative Colitis since he was 25 years old with a flare up in the last three months. He's in the bathroom thirty times a day with loose, urgent, bloody stools. This happens every three to four years. He has fatigue with loss of blood. He has night sweats since receiving enemas with steroids two weeks on and two weeks off. His eyes are red and blurry. He eats a low fiber diet watches carbs, eats cheese, almond milk, some fruit, meat, potatoes, drinks beer seven times a week, chews tobacco, eats some sugar products three times a week. He had mono as a kid and is sensitive to wheat. He takes Azulfidine for the Ulcerative Colitis. **Treatment:** Acupuncture to bring up the immune system and balance the bowels. ITM herbs to cool the heat of the Ulcerative Colitis irritation and build blood.

Analysis: After four treatments spread two weeks apart he had gained ten pounds, energy was better, occasional bouts of loose stools, NAET would help this person long term with underlying food allergies. Beer and chewing tobacco can aggravate Ulcerative Colitis.

Diverticulitis: In Chinese Medicine it is treated similar to Ulcerative Colitis. Treat for parasites, food allergy and immune depression.

Synopsis: Sixty cases of constipation usually secondary to other problems, usually digestive. The digestion needs to break down the food to produce the nutrients for other body functions. Osteoporosis occurs due to no absorption or breakdown of food. Alzheimer's occurs due to lack of endorphins from fat breakdown. Low energy, fatigue, depression, and thyroid, slows down the bowel movement. Pain due to impaction of bowels contributes to headaches and depression. The energy in the body needs to **MOVE,** constriction in any organ starts the breakdown process. Hemorrhoids occur due to slow bowel and stagnant liver. Arthritis occurs due to unprocessed food stuck in joints. Eczema and psoriasis occurs due to toxicity of unprocessed food. Allergies can create the constipation. Hernia occurs due to straining. Prostate congestion occurs due to toxic colon due to constipation. Foggy mind occurs due to toxicity from stagnation of the colon. Stagnation creates a hot bed for viruses, infections, multiple ear infections, colds, pneumonia, and lung congestion. Lungs and large intestine are related in Chinese Medicine so an issue in one area usually leads to an issue in the other. Menstrual issues from stagnation can occur such as pain, endometriosis, fibroids and ovarian cysts.

Bunions

People who have bunions suffer from an enlargement of the big toe joint. If it is only pain and on the right big toe it is gout. Gout can also show up with redness at the joint. Bunions are usually whitish and can create a lot of pain. Acupuncture can help. I have had several people with one treatment that had the bunion disappear, others we treated with whatever the major complaint was and the pain receded within two to four treatments. Bunions are the points for the lungs on the feet in reflexology, homeopathically related to tuberculosis in the family's past.

Cancer

58 treated, 33 females, 25 males, average age 64 years old.

Chinese Medicine works well as a standalone or immune support for people who are going through any chemotherapy, radiation or medication for cancer. In New Mexico Doctors of Oriental Medicine were considered Primary Care Physicians; we could order blood tests and x rays. One third of my practice worked with cancer clients.

Acupuncture supports the immune system, white blood cells, while detoxing the immune system from the offending toxins, viruses, parasites, tangles whatever is causing the cancer. The procedure is a process of support and cleanse. Care needs to be observed of support and cleanse. Care needs to be observed to not push the client farther than their immune system is able. The clients, once cancer is achieved have a very weak immune system in the 4th and 5th layer of illness. (See Immune System).

Protocol is acupuncture 1 time a week and ITM Chihko Curcuma (turmeric) cancer scavenging herbal combination. We then added an herbal for local cancer issue: ITM herb Lindera for Prostate or Uterine (endometriosis). ITM herb Blue Citrus for breasts. ITM herb Sparganum for lumps/fibroids. ITM herb, Laminana for lymphatic swelling, cysts (Non-Hodgkin's/Lymphoma). ITM herb Belamcanda for lung tumors, ITM herb, Paris for metastasized tumors.

Acupuncture helps with the pain from radiation and chemotherapy, hot sweats due to medication, urinary incontinence, shortness of breath, circulation of lymph from swelling, thyroid support, healing of scars, and foggy brain from chemotherapy.

In the 50's and 60's women were given a lot of Hysterectomies. These women were then put on hormone replacement therapy. This estrogen support caused estrogen sensitive breast cancer in many women. There is more discernment in therapy now with hormone replacement. Chinese Medicine herbals provide the pieces to make the hormone if needed in the body.

Post cancer care includes a good diet of rainbow foods, little alcohol, caffeine or sugar. Regular immune care, for example; massage, acupuncture, yoga, meditation, chemically and toxin/mold free environment, exercise, surround self with happy people. There is an emotional connect with cancer and organ affected. (See Emotions)

Overabundance of parasites in lymphatic system can cause Lymphomas Leukemia's or brings the immune system down enough to allow other issues into the body.

Neuropathy Case Study: A 67 year old female presented with lymphoma one year ago. She did chemotherapy with resulting neuropathy in her hands (tingling, burning) and it sometimes affected her feet. Her energy is low and she has low back pain. Her neck and shoulder hurt and she has a subclavian port. Medications: aspirin, acyclovir, fluconazole, lorazepam, amitriptyline, and docusate. She has pollen and mold allergies. She eats a balanced diet with no sugar and has caffeine three times a day. **Treatment:** Acupuncture

Analysis: After three treatments she reported having better energy, neck and shoulder was better and not as tight, low back felt better and the neuropathy was mostly resolved in the hands. She had other procedures coming up so I did not see her again. Caffeine can create more heat in the system aggravating neuropathy. One dose of caffeine per day is okay. The chemotherapy created the constipation, therefore the docusate. The chemotherapy created the neuropathy.

Kidney Case Study: A 55 year old female presented with back pain due to kidney surgery for cancer on the left and half of the right kidney removed. Spots found on the lungs, sinus pain, low energy, panic attacks, high blood pressure due to pain, diagnosed with Addison's and on steroids, kidneys at 50% functioning. She was very hot, had migraines 3-4 times a month. She had dry eyes, shortness of breath with a cough, occasional heart palpations. She had IBS in the past with a lot of belching and gas. She was up at night 3-4 times to urinate. She had mono in her twenties. She craves sugar and drinks power aide plus diet mountain dew. She had complete hysterectomy because of cancer. **Treatment:** Acupuncture, ITM herbs, NAET

Analysis: Her history indicates that she did not absorb her nutrition with a diagnosis of IBS. Her diet was not good, a lot of sugar and drank diet mountain dew for years in large amounts daily. Diet mountain dew contains aspartamine as a sugar substitute. It is a carcinogenic, that means it created cancer in mice studies. She had 1-1/2 kidneys removed because of cancer. The cancer had now metastasized to her lungs. She was in a lot of pain. We used ITM herb Chihko Curcuma to cleanse the cancer out of the cells. ITM herb Restorative to rejuvenate what we could of the kidneys for more energy and to build blood. We used ITM Belamcadae for sinus and coughing, Echinacea herbs for canker sores. She used trace minerals for charley horses in her legs. Acupuncture for immune system, sinus, pain, vertigo, digestive issues, itching and shortness of breath. We cleared out gall stones with ITM herb Lysmchia that softened the stones so she could excrete them without pain. We began NAET so she could absorb her nutrients better. She improved, having less constant pain, eating better, improved energy and no itching. She was at end stage cancer. We were able to extend her life for a year and a half with less pain and better quality of life.

Lung Case Study: A 76 year old male presented with lung cancer and 1 lower lobe of lung removed in 1993. He was diagnosed with metastasized nodules in other lung. He refuses chemo or experimental drugs. He had a tumor removed in the rectum in 1992 with radiation therapy. No medications at this time. He eats fried foods and Mexican food and 2 cups a coffee a day. He walks a lot. He has a colostomy bag because of previous surgery. He is worried.
Treatment: Acupuncture, ITM herbs
 Analysis: He is put on ITM herb Paris for metastasized tumors and ITM Belamcadae for the lungs. He had 8 acupuncture treatments to boost the immune system and lungs plus ITM Laminara for lymphatics. 5 years later he was still alive and outlived his wife.

Chronic Lymphocytic Leukemia Case Study: A 56 year old male presented with shingles, diagnosed with hypoglycemia, quit smoking 1 ½ months ago, quit drinking 3 years ago. He had mono in college, he tends to be warm, cold sweats at night, headaches and sometimes dizzy. His body aches and he get cold sores. He has trouble sleeping and has loose stools. Further testing resulted in high lymphocytic count.
Treatment: Acupuncture, ITM herbs

Analysis: When this client came to me he wasn't sure what was wrong with him. He was diagnosed 2 months later with Chronic Lymphocytic Leukemia (LLC). Acupuncture was used to boost the immune system and he started feeling better. He took ITM herbs, Antler and Gymnostemma for energy and to support a deficient immune system. He had worked in a Uranium mine with no protection from radiation for 2 years in 1975-1977; all the other workers from that project were deceased. He had nodules in 1991 in his armpits and groin plus itching all over. He was working at another copper mine at time of death in 2002. He was negative for HIV. Towards the end we watched his white blood count elevate. He broke out in herpes at the end when his immune system was out of control with a lot of pain. He had staph infections, parasite issues often. Took chemotherapy in 1999 to back off the Lymphocytic with no success. Mosquito bites would turn into infection. He did 130 food tests for allergens and followed his blood type A diet, which helped him feel better. Towards the end he had bowel trouble due to Vicodin with codeine in it. He was achy and fatigued; the body was working hard to produce all those white blood counts. His spleen enlarged.

Chinese Medicine provided him with immune support in the beginning and comfort care at the end, his life was extended 8 years to see his daughter graduate, marry and have a child.

Chemical Sensitivity

Chemical Sensitivity can precipitate from immune dysfunction, allergies, virus, candida, autoimmune disease, fibromyalgia, too much medication, Alzheimer's, old age, exposure to environmental problems like sick buildings, pesticides, preservatives, amalgams, vaccines and spider bites.

Toxic Exposure Case Study: A 53 year old male presented with toxic exposure to carbon monoxide while working in a small exhaust area in construction work. Balance/Vertigo issues he has to crawl if there is no light. He has peripheral floaters and an issue with swallowing. He disconnects and flattens out emotionally, no energy, depressed immune system, and migraines worse with no focus. He has ringing ears and is difficult to recover from getting sick, had Epilepsy as a child. He was taking glucosamine/chondroitin, vitamin C, Vitamin B, fish oil, Benicar for blood pressure. **Treatment:** Acupuncture and ITM herbs.

Analysis: Acupuncture can slowly detox a body. We put him on Gecko and Restorative herbs to boost both kidneys for energy and immune system. We did acupuncture for vertigo/balance issues and points for swallowing improvement. Clarity was an issue that the carbon monoxide dampened in brain function. Acupuncture opens up the circulation to the brain. The herbs supported kidney function, which also charges up the brain. After three months his head stayed clear, had less headaches and no vertigo. After nine months the carbon monoxide issue was fairly resolved. He then started dealing with his depressed immune system.

Cholesterol

The American diet in general is not good. Too much sugar, too many carbohydrates that are turned into glycogen creates a fatty liver. The cholesterol rate goes up. Green vegetables are good for the liver and should be eaten at least 2-3 times per day. Good fats also help the liver and the body for dry crepey skin, dry eyes, making endorphins for the brain, steroids for allergies and body pain. Good fats include butter, and olive oil, coconut oil, try to eat no GMO products or pesticide/preservatives and eat as clean as possible. Normal cholesterol rate 15 years ago was 250. If one goes too low on the cholesterol rate one ends up losing the benefits of good healthy fat. **Treatment:** Acupuncture used to increase metabolism of the liver and digestive system; stimulates the body to break down the cholesterol naturally. Herbs are used to reduce high cholesterol in the bodies system as well. I have had clients on herbs for years with no negative results. Change of diet is most important for long term success. (See Hepatitis C case)

Concussion

TBI-Traumatic Brain Injury: Concussion is usually an unseen issue and hasn't been addressed much until recently. TBI or Traumatic Brain Injury is a bruising of the brain. Progressive injury creates degenerating conditions of brain fog, brain fatigue, spacey, vertigo, balance issues, and cognitive acuity degeneration.

Chinese Medicine can recharge the brain through Acupuncture and herbs (ITM Bu Nao Wan, and others). I've had cases of football, car accidents, and Veterans of War. It was slow progress to regenerate the older cases. These cases can be treated like someone who has had a stroke (see cases) It took at least a year with acupuncture every two weeks. Acute cases 1-2 months.

Concussion can be monitored by dilation of the pupil in the eyes. As the client gets better the eyes dilate less often. When the eyes dilate the client should rest, it indicates that the brain is overwhelmed.

Depression

In Chinese Medicine depression relates to a congested liver due to toxicity or stagnation.

Stagnation relates to emotionally not being able to move through situations. The Chinese belief system indicates a cycle of beginning, journey, completion and death to be redone over and over again. Toxicity can be created by a number of things malabsorption of food, allergies, overdose of alcohol or drugs, legal and illegal, or exposure to a toxic environment. If the body is not processing properly the toxicity can lead to stagnation. **Treatment:** Acupuncture moves the stagnation and helps to detox toxicity. Depending on how stagnant a person is determines the treatment schedule. Technically, twelve treatments for one issue in Chinese Medicine, especially emotional. It takes time to change the behavior or habits related to the break in the flow of life. Herbs are also good for chemically detoxing and promoting flow of energy.

Diabetes-Type II

59 clients were treated for diabetes, 39 females and 29 males at an average age of 57. 90% of these clients improved with diet change and NAET. Many were able to reduce their medications for Type II Diabetes. Blood sugar reduced for Type I.

Diabetes Case Study: An 80 year old male presents with Type II diabetes for ten years. Urinating every two hours at night, fatigue, blood sugar between 170-190, ears tend to get plugged, with possible gluten intolerance. He watches diet, but likes sweets and carbohydrates. **Treatment:** Acupuncture, ITM herbs, NAET.

Analysis: Acupuncture and ITM herbs for energy-Man's Treasure and ITM herb Gecko. Started treating one time a week for one month, then every two weeks for six months. We then did one treatment for maintenance every three weeks. Acupuncture is used for the kidneys to better support the digestion of food. NAET is used to correct the underlying food allergy. Diet change is necessary to get proper nutrients into the system. The blood sugar dropped to a range between 113-130's when client was compliant on diet. His energy increased with less frequent urination at night.

45

Compliance of diet becomes an issue because most diabetics love the sugar and carbohydrates resulting in not eating a proper diet. Cholesterol levels can run higher because these people prefer carbohydrates like bread and pasta over vegetables and protein. Diet for diabetes consists of kidney foods; protein, especially meat, root veggies like beets, potatoes or carrots, blue food like blueberries and blackberries. Bad for the diet is sugar, caffeine and alcohol. One dose of each a day is okay, but to go to green tea with pycnogenol is better and fruit is okay within reason. Alcohol fermented is worse for the body like beer and wine.

Stress will raise sugar counts and so will infection. Diet compliance is necessary to maintain a low blood sugar. Consistent low sugar shows up in A1c reports. Fatigue occurs when client is not absorbing the nutrients from the food or poor diet.

Food allergy, intolerance or malabsorption is usually a factor. Food allergy is picked up in blood tests but, intolerance and malabsorption are usually only picked up with kinesiology. NAET can help greatly to facilitate reabsorption of necessary nutrients.

Digestive Issues

Digestive issues are the primary cause of a host of physical problems in the body. This is one of the schools in Chinese Medicine. "Correct the Digestion and everything works". The issues addressed are acid reflux, heartburn, hiatal hernia, ulcers, heat in the stomach, bloating, belching, swelling, (cold kidney/digestion), yeast/candida overgrowth, constipation, gas, diarrhea, parasites, eczema, psoriasis, cramping, nausea, thyroid low slows food down, thyroid high can't eat enough, allergy-sinus, chronic congestion, fatigue, depressed immune system. **Treatment:** Acupuncture can charge up the bodies system (kidneys) so a person can digest their food better. Neutralize the temperature in the body can correct many issues. NAET to correct allergy issues can help correct Ph issues; acid/base.

Candida: is an overgrowth of yeast in the system that actually can change hormonally what is going on in the body. It creates craving for sugar and carbohydrates to feed it. It takes seven years to change all the cells in the body. It appears to be very addictive and hard to get rid of like if one is smoking, does alcohol, drugs or hormonal replacement. It is usually the core reason why someone can't lose weight. If a person is not getting better with their digestion, yeast is probably the underlying issue. (See Tangles) See Dr. Crook's "Yeast Connection".

Leaky Gut: occurs because parasites have eaten holes in the bowel system and food nutrients are found were they shouldn't be which sets up an allergy reaction-autoimmune response. (IgA, IgD)

Weight Loss: correcting core digestive issues with acupuncture charging up the system so the food cooks and breaks down properly into usable nutrients. Correct diet so proper nutrition feeds the body. Herbs to correct heat or cold problems, like probiotics. NAET can help with absorption issues due to food allergy. The body needs maintenance just like a car, if you don't have good gasoline it knocks!

Candida Case Study: A 20 year old female presents with kidneys hurting, chronic acne, diagnosed with fibromyalgia, fatigue, large intestine slow after appendix removal-has to massage the area for food to go through. History includes mono, herpes on hands, lips and sometimes forehead. She's been sick for two years. MMR vaccine gave her high fevers of 105 one time a month. She's allergic to food, pollen, mold, wheat and dairy. Her body presents hot with cold hands and feet, blurry, red, dry eyes, no appetite, insomnia, irregular menses with pain and a lot of clotting. She tries to follow a sugar and yeast free diet. **Treatment:** Acupuncture, ITM herbs, homeopathy, diet-yeast free.

Analysis: Through kinesiology she was diagnosed with four viruses, mono being the gateway virus as a child. She had herpes virus, shingles, strep in the lungs and a retro virus, parasites-lymph flukes, and a pesticide/formaldehyde toxicity including an issue with mercury. She lived some hours away so the homeopathic were staged one per month. She was put on a strict yeast free diet. She got all the mercury out of her teeth which can feed the yeast and viruses. She was taking digestive enzymes. After five years she was doing well with school and a relationship.

It took determination to get through the gentle cleansing with the homeopathic and then the building up of the immune system afterwards. She inherited severe allergies from her father, which depressed the immune system and then the viruses jumped on.

GERD Case Study: A 48 year old female presented with gastroesophageal reflux (acid reflux), hypothyroid, cold hands and feet, occasional night sweats, neck and shoulder pain, and frustrated. She eats gluten free, nutritious shakes, veggies, meat and no milk products. She had a bad case of chicken pox as a child and a lot of polio shots. She has cold sores often. **Treatment:** Acupuncture, ITM herbs, homeopathy, NAET

Analysis: Acupuncture was used to warm up her interior to digest her food. Acupuncture used to clear heat at neck and shoulders related to frustration. She was menopausal, which causes heat in the system when the adrenals take over the estrogen function for the ovaries. That internal heat can cause GERD. She also had food allergies that we did NAET for. After a year her reflux was not an issue, but the menopause/endometriosis became the underlying cause.

I treated 75 clients with digestive issues. Of the 75, 41 were female and 34 were male, and the average age was 46 years old. I treated 19 clients with candida/yeast conditions with an average age of 44 were all female. The sensitive women would complain that their partners would give them yeast through intercourse. For the females health, compliance of a yeast free diet by the male would help their partner.

Ears

Allergy Case Study: 49 year old female presented with allergies later in life, eyes itchy, ear full of fluids, fatigue, insomnia, cough and drainage. Diagnosed with diverticulitis, she presented with cold hands and feet, blurry itchy eyes, ears popping, constipation, belching and gas. Her diet included meat, veggies, fruit, milk, cheese, 2-3 cups of coffee, a lot of salt, and sweets one time a day. Her menses was regular. **Treatment:** Acupuncture, ITM herbs, NAET, Diet

Analysis: After one treatment her nose opened up and she had more energy with the herbs and acupuncture. She proceeded through the NAET allergy elimination program, 14 treatments. By the end she had no issues with sinus, fatigue, insomnia, cough or drainage. Her bowels were working well, no itching eyes or popping ears. She craved salt, which indicated that the kidneys needed support. Her caffeine was reduced to clear excess heat in the system resulting better sleep and less constipation.

Ears Ringing Case Study: A 60 year old male presented with ears ringing constantly, high pitch, wax cleared out and no change, fatigue, insomnia, and shoulder blade pain. He is a heavy equipment operator. **Treatment:** Acupuncture, ITM herbs

Analysis: Ears relate to the kidneys in Chinese Medicine. ITM Restorative herbs for left kidney and ITM Man's Treasure herb for the right kidney was given to boost energy to the kidneys and the ears. After seven consecutive treatments done one time a week for the right shoulder and the ears his right ear was immensely better and his left was improved after six treatments with no ringing.

Ears can improve greatly with acupuncture and herbs by tonifying the kidneys. At least six treatments are necessary to improve, if not more. Damage to the inner ear usually does not recover from a concussion, whiplash, but it's worth a try to treat. I treated fifty four clients for the ears, of which thirty three were female and twenty one were male. Important to check for tooth issues, viruses, allergies/sinus infections from food/pollen, TMJ, trigeminal neuralgia, shingles and migraines.

Emotions

Each organ in Chinese Medicine relates to an emotion. When taking an intake, it is helpful to discover the primary emotion of an individual. This can indicate which primary organ is deficient or excess. The question is does the emotion create the physical problem or vice versa? Usually it's a combination of both. **Liver:** Anger-excess, Depression-deficient **GallBladder:** Overwhelmed-excess, irresponsible-deficient **Stomach:** Worry-excess, focused obsession-deficient **Spleen:** Empathy/sympathy-excess, lack of-deficient **Heart:** Joy-excess, Anxiety-deficient, **Small Intestine:** Ability to process, separate what to keep and what to let go **Large Intestine:** Elimination-ability to let go **Kidney:** Fearless-excess, Fear-deficient **Pericardium:** Heart protector-ability for self-care **Triple Warmer:** Lymphatics-ability to move through situations.

Eyes

Treated 60 people, 39 Females, 21 Males and average age of 57 years old. Eyes relate to the liver in Chinese Medicine. Whatever affects the liver affects the eyes. Heat is the biggest culprit because when the liver is traumatized it heats up. Anger, stress, too much chemical load from a toxic environment, pain medicine, allergies, viruses, acid ph, drugs, sugar, alcohol, caffeine all heat up the liver. Acupuncture can cool the body/eyes, diet change helps with a more alkaline diet, good oils-olive or coconut keep the moisture in. There are great herbal combinations for retinitis, glaucoma, macular degeneration, cataracts, and eye infections. The eyes have been known to improve in vision with herbs and acupuncture.

Eye Allergy Case Study: A 55 year old male presented with blurry, red eyes year round due to allergies. He works as a livestock auctioneer. He was fatigued, sweats easily, body is hot, and he's light headed. He had migraines when younger, itchy ears, constipation, insomnia. He had allergy shots for bread, melons, chlorine, and beer. He takes Zyrtec D. **Treatment:** Acupuncture, NAET

Analysis: He was treated for pollen allergies and barn animal allergies. Felt better after the NAETS; no further trouble. Too much heat in the body caused his constipation, sweats, insomnia and migraines. Reducing allergic reaction neutralizes the heat in the body and reduces sweat.

Dry Eyes Case Study: A 63 year old female presented with chronic red, dry eyes for 15 years, stress, anxiety, diagnosed with scleroderma, hypothyroid, diverticulosis, she continually had UTI's, and taking a lot of antibiotics. She presents with cold body and hands and has night sweats, gets shortness of breath with anxiety, urinates frequently, and fatigued. Her diet is good, but too much coffee that creates heat in the system contributing to red dry eyes, anxiety, bowel issues, and UTI's. She is taking Cipro and Bactrim for the UTI's, Levothyroxine for thyroid, eye drops, Vitamin C, D, Omega 3, Omeprazole for digestion, and Ibuprofen PM. **Treatment:** Acupuncture, ITM herbs, homeopathy.

Analysis: She was put on ITM Restorative herbs to clear heat in the body for the UTI's and dry eyes. She was instructed to use olive oil instead of lotion on her body after showering; the oil holds moisture in. After three treatments her dry eyes of 15 years were doing well without her medication or eye drops. She attributed it to the using the olive oil. The acupuncture cleared the heat in the body. She was put on ITM Pyrosia herb for chronic UTI for a year. She was able to stop the antibiotics. She had several viruses, mercury and chemical toxicity that we removed through homeopathy. These viruses and chemicals create unwanted heat in the immune system that is constantly trying to get rid of the antigen/antagonists. The homeopathy destressed the hyped up autoimmune system. (Scleroderma)

Fatigue

Fatigue is one of the primary problems why people show up at my office. This is a kidney issue. Left kidney is the will and Right kidney is the will to carry it out or complete an action. (See Kidney/Diet/Brain Tangle) The body runs on electricity and when the batteries "the kidneys" are too low the body starts to malfunction at the weakest place for that individual. There could be a genetic predisposition, allergies, bad diet or bad habits. There could be stress or emotional issues, anxiety long term can crash the body and cause fatigue. Pain causes chronic fatigue.

Adrenal Exhaustion Case Study : A 22 year old female presented with shaking for years and now it's getting worse. She was under extreme stress at college. She had constipation, extreme fatigue, and eyes dilated. She had a cold body, hands and feet, night sweats on and off, urinated frequently, anxious, and worried. Her diet was composed of junk food, little veggies, occasional fruit, carbohydrates, milk, cheese, sugar products and one soda pop at least once a day. She had mono at 4 years old and all her childhood vaccines. **Treatment:** Acupuncture, ITM herb Restorative.

Analysis: Acupuncture to improve energy and clear heat from stress. Heat showed up as constipation and night sweats. Decreased kidney energy shows up as shaking and dilated pupils. Both kidneys needed help because of mixed symptoms of heat and cold. She had progressed from adrenal fatigue to adrenal exhaustion. Increased urination due to kidneys letting go and not able to control. The junk food diet did not help her case and she had mono as a 4 year old-a gateway virus suppressing her immune system. Adrenal exhaustion can take years to reboot with careful diet, acupuncture, herbs and monitoring her energy output on a daily basis.

Fibromyalgia

Fibromyalgia is a complex chronic disease that affects mostly women. In Chinese Medicine it is related to the liver, which rules muscles and joints. Acupuncture can greatly reduce the pain and increase circulation. There can be an underlying chemical or food allergy, virus load, or autoimmune disease. I've treated 30 people, 25 female and 5 males average age of 47 years old. All have some form of serious issues underneath the fibromyalgia, too many vaccines, kidney removed, molested, extreme emotional duress, hyperthyroid, Hep C, arthritis and psoriasis.

Fibromyalgia Case Study: A 57 year old female presented with all over pain in legs, head, shins, incontinence, headaches, constipation, fatigue, insomnia, sadness, runs hot and cold, ringing ears, irregular diet includes soda pop one time a day, irregular menses with pain and clotting, finished menopause last year. She takes Haldol one time a month PRN. She has short term memory loss, hears voices and humming. **Treatment:** Acupuncture, herbs, homeopathy, ITM Restorative and Woman's treasure

Analysis: She was treated for six months one time a week. At four months her energy was better, less pain in legs and hips. Removed with homeopathy Coxsackie virus that causes pain in hips and thighs. As kidneys improved the voices receded related to brain function. She had been in the state hospital a few times, given a lot of prescription drugs and poor diet did not get the nutrition to the brain. The heat would cause headaches, constipation and insomnia. The cold encourages incontinence and all the drugs to sedate in hospital caused a lot of heat in the body generating fibromyalgia from the toxicity.

Foot Pain

Foot pain: Includes heel spurs that stabilize the bone structure in the foot when a person is walking too much on their feet. It takes 6-12 acupuncture treatments to recover the heel and then maintenance treatment every six months, sometimes it corrects and no more problems.

Plantar Fasciitis: The tendons between the toes on the bottom of the foot become painful. This is the stretching of the ligaments that attach to the muscles in the calf. Acupuncture on the calf muscles usually release the tightness connected to the feet. Four to six treatments one time a week sometimes up to twelve treatments is needed.

Hammertoe: Acupuncture at the joints can sometimes correct the issue, a 50-50 chance and worth a try. Prognosis not good if surgery has been done, but can relieve the pain. I've treated 27 people, 20 female and 7 males at an average age of 47.

Foot Case Study: A 57 year old male presented with Raynaud's, neuropathy in hands and feet, arthritis, frequent urination, enlarged prostate. He tends to be cold, dry eyes, Type A diet. **Treatment:** Acupuncture, ITM herbs, homeopathy

Analysis: After four treatments one time a week with acupuncture the neuropathy in the feet reduced, after six months there was no more complaint. Homeopathy was given for viruses, chemical toxicity and parasites. After five months the arthritic hands (Raynaud's) was better, energy good, urination less frequent. ITM Restorative herb was used to improve the urination, kidneys and dry eyes. Raynaud's is a circulation issue in Chinese Medicine. We unconstrained the Liver Qi so blood flows better through the hands and the body. Extreme stress blocks circulation and can cause neuropathy.

Gout

Gout is the inflammation of the right big toe joint. It is a chemistry issue. The body is not metabolizing purine foods, alcohol, some fish, shellfish, seafood, anchovies, sardines, bacon, turkey, veal, and organ meats. Treated 10 people, 2 females, 8 males; average age of 57 year old.

Gout Case Study: A 44 year old male presents with big toe continually numb. His left hand fingers and thumb are numb off and on, progressing after last six months. He has pain between the shoulder blades. He had a car accident in high school and chipped his left elbow. He had thyroid lumpectomy right side and appendix out. Tends to be warm and snores all night, not rested. Excess belching, dribbles with urination and can't shut it off, craves sugar, beer every day, pot of coffee and diet coke every day. He eats meat, milk products and carbohydrates. He quit smoking one year ago. He does carpentry work. **Treatment:** Acupuncture, Diet

Analysis: He is eating and drinking all the wrong things that create gout (purine foods). After three treatments and change of diet there was no more big toe pain. After five treatments the shoulder pain and hand numbness resolved. He was sleeping better and his energy was good.

Hands

Hands: Trauma most often is the issue or repetitive use. Pain can be generated from whiplash, concussion, thoracic outlet, carpel tunnel, arthritis due to food allergies, and circulation from heart issues, diabetes and neuropathy. Acupuncture primarily is used to relax the muscles, nutrition to stagnant muscles helps; for example trace minerals, NAET can help food allergy issues and absorption of nutrients. Homeopathy can remove virus and chemical restrictors in the tissue. Of 53 people treated for hands, 37 were female, 16 male and average age of 57.

Trigger finger can be helped sometimes by relaxing the muscle that holds the finger static. It is worth trying acupuncture and we'll know within 3 treatments.

Hands Case Study: A 55 year old massage therapist broke her wrist and has been in a brace for five and half weeks. Her thumb and second and third fingers are numb. It is worse at night when she is tired. She's a Type 1 diabetic since she was 11, had her gallbladder out, and a C section. She had a tendency to be warm. She follows a fairly good diabetic diet. She does have strong coffee once a day and sometimes a diet soda. **Treatment:** Acupuncture, electro-stimulation, castor oil

  Analysis: Diabetics have a harder time healing wounds and bones. Electro-stimulation was used around the hand (carpal tunnel) and the shoulder (Thoracic outlet syndrome) to relax the muscles around the nerves. These muscles were creating compression, therefore numbness. The needles actually reset the muscle to a looser configuration to reduce pain. Massage and chiropractic stretch the muscles to open up the nerves to reduce pain, like a rubber band. For long term benefits the resetting stabilizes the muscle open. Massage and chiropractic work very well with acupuncture, each enhancing the other. Castor oil reduces inflammation and swelling. After five treatments the thumb and fingers were significantly better. She was sleeping well. She reduced her caffeine intake, reducing the heat in her body, which kept her awake and could also cause more pain. Diet products with aspartamine were removed. (cancer causing)

Carpal Tunnel Case Study: A 58 year old female presented with chronic carpal tunnel syndrome from long term over use as a court stenographer. The arms and hands swell, thumbs feel like they can fall off, numbness in both the fourth and fifth fingers with numbness. She has restless legs, digestion issues, herpes, allergies, shortness of breath with allergies to pollen and possible food. She is through menopause. She does weight watchers diet. She has whiplash from a car accident. **Treatment:** Acupuncture, ITM herbs, NAET, homeopathy

Analysis: This person needed to be staged in working with her health. The pain/whiplash came first. She used liquid trace minerals, a cap full every night in juice. Restless legs resolved within two weeks. Fourth and fifth finger numbness indicated C4/5 compression from the car accident. Acupuncture one time a week for neck and arm pain resolved within three months. Allergies addressed next with some colds in between. One always has to treat colds and flu first before addressing deeper issues in Chinese Medicine. Allergy can flip a person into colds and flu, often because it depresses the immune system. She had a lot of allergies and it took six months to treat with NAET. Homeopathically she was treated for mercury and pesticides in the tissue, viruses and black widow, which can cause constriction in the muscles and nerves. Black widow bites can remain in the body over time and add to multiple underlying issues. She is now able to travel and be productive after four years of treatments. She still needs maintenance, especially for whiplash.

Headaches

There are five kinds of headaches in Chinese Medicine. Frontal relates to digestion, sides of head (temporal) relates to gall bladder, top of head relates to heat/over thinking, occipital/neck usually relates to trauma/whiplash, the fifth is an internal heat, relates to foggy brain-kidney problems, migraines. I treated 122 individuals for headaches; 86 were female, 36 males and average age of 42.

History of Hepatitis can aggravate headaches, epilepsy, seizures, stenosis of cervical vertebrae, hyperthyroid, high blood pressure, fibromyalgia, TMJ, anxiety, stress, emotional constraint, allergies, menopause, menses, chronic sinus, whip lash, concussion, ear infections, low back pain, shoulder pain radiating upward, carpal tunnel, pregnancy, anemia, heat exhaustion, low energy, fatigue after child's birth.

I treated 69 individuals for migraines and 64 of them were women and 5 men and the average age was 43. There is internal heat present that rises to the head to be released, the underlying issue needs to be addressed.

Migraines/Headaches Case Study: A 40 year old female presents with migraines and headaches on right side, anxiety, right shoulder blade tension, seasonal allergies, ears itch, drainage back of throat, fatigue, tends to be cold with hot sweats at night, fairly good diet, 2 cups of coffee in the morning, craves sugar, but curbs it. She has normal menses with heavy first day. She had mono in high school and has possible autoimmune now. She can get hives when hot and is on hormone replacement therapy. She takes multi vitamins, probiotics, vitamin D, Omega 3's. **Treatment:** Acupuncture, ITM herbs, NAET

Analysis: She has gallbladder headaches affecting the temporal side of her head; migraines. She has a history of stress and anxiety. Allergies heat up the body's immune system to protect itself. Chronic allergies can lead to autoimmune disease, if a person never gets a break from the irritation. Gallbladder relates to being overwhelmed, when liver/gallbladder heat up together, the heat rises to the head to be dispersed ending in a headache. The kidney's job is to cool the liver/gallbladder. If liver/gallbladder are overwhelmed by constant allergies, the kidneys cannot calm them resulting in headaches; underlying all of this is a premenopausal woman. This also drains the kidney energy because the ovaries are quitting their job and giving it to the adrenals/kidneys. The kidneys were already overworked due to allergies. She had a history of viruses, chemicals and parasites. We started NAET to treat the allergies to get the load off her immune system. At the end of treating the allergies she had no ear itching, no sinus congestion, anxiety was improved, energy was better, no migraines, except on and off with the beginning of her menses. She ended up getting a hysterectomy, which ended the occasional migraine. She had no problems after that.

Sinus headache Case Study: A 34 year old female presented with sinus headaches her whole life, was worse in the winter and mid back pain. She has a dull headache everyday with sinus swelling, eats a good diet, craves sugar at menses, had all her childhood vaccines. **Treatment:** Acupuncture, ITM herbs, NAET, homeopathy

Analysis: Stared with acupuncture and ITM herbs to get the infection out of the sinuses and to give her more energy (ITM herb Woman's Treasure). Started virus removal with homeopathy for the chronic infection. Did NAET to remove allergies over a year one to two times a month as her budget would allow. After two years she was stable, unless she identified a new allergen for herself. No reports of further sinus headaches, had more energy and no back pain. Mid-back pain relates to liver/gallbladder points in Chinese Medicine and can be aggravated by too many allergies to process over time presenting in chronic back pain.

Heart Conditions

High Blood Pressure, circulation, heart palpitations, stroke, TIA's (Transient Ischemic Attacks), 14 people treated of which 7 were female and 7 were male; average age was 56. **NIH** (National Institute of Health) did a double blind study during the Clinton years that showed acupuncture can reduce blood pressure significantly. I have had quite a few people come in regarding that and was able to help them reduce their blood pressure, correct diet, increase exercise and extend their life. Some were on pacemakers, had angioplasties, chest tightness, heart palpitations. Underlying issues of heart are a fired up liver that overacts on the heart. This can happen because of stress, anxiety, emotional issues, caffeine, sugar, alcohol, drugs, poor diet, anything that heats up the liver. If the kidneys are not cooling, the liver then the liver overreacts on the heart. I have seen blood pressure come down from 210/130 to 160/109 after six treatments and maintain for a year. Diet and stress level must change. Normal is 120/80 in your 20's and 30's, normal after 40 is 140/90. As a person ages the circulation system becomes more open with use. It takes more pressure in a loose system to get the blood around the body. Older people on too much medication can cause vertigo, fatigue and increased falling because they have no blood pressure in their system. Circulation is increased with acupuncture and herbs by relaxing the circulation system.

Treated 22 strokes, 13 female and 9 male; average age of 69 years old. Acupuncture for stroke helps with balance, dizziness, speech impediments, face droop, grip for hands, ambulation for legs, tightness of chest, circulation in extremities, forgetfulness, numbness of extremities, neuropathy in extremities, swelling legs and face, fatigue, reduce mini-strokes, increase strength to extremities. The sooner one treats the better the results.

Aortic Stenosis Case Study: A 65 year old male presented with diagnosis EKG Aortic stenosis, heart palpitations for one month, with calcium build up on valves, palpitations when lying down, night sweats, shortness of breath, urination slow, left ankle swelling at night, diet fair, heavy on cheese and yogurt, caffeine two times a day, alcohol two to three times a night, and hard time sleeping. **Treatment:** Acupuncture, ITM herbs, Restorative, Vitality.

Analysis: After three months on the herbs and four acupuncture treatments he was feeling good, more endurance, more energy, better urination, no heart palpitations, no night sweats or heartburn. The Restorative cleared the liver heat that was firing him up at night, sweats, insomnia and heart palpitations. Vitality herbs helps the heart beat better reducing ankle edema. There was still a little shortness of breath. He corrected his diet and started exercising more.

Stroke Case Study: An 86 year old male presents with left sided moderate to severe Ischemic Stroke diagnosed in the emergency room. Signs and symptoms were left face numbness, left side eye and lips drooping slightly, left thumb and index fingers curling, numb tips of fingers, left hip achy, general thighs weak; progresses within one hour to left side face droop, left lips, cheek and eye lid droop. Pulse fluttery, no tongue deviation, blood sugar is 159; was observed by practitioner, Dr. Ragle in meltdown. **Treatment:** large doses of aspirin given and sent to hospital; 48 hours later was released from rehab/hospital.

While in the hospital admittance area the practitioner pressed points along gallbladder and bladder meridian to release pressure of stroke on both sides of the head. **Treatment:** Acupuncture one time a week for one month, gingko herbs for brain function, acupressure, aspirin. One week later some left side vision was afflicted and brain not able to concentrate. One month later was able to do taxes, eyes better, some numbness in lips when tired drooping occurs, recovers with rest. **Analysis:** Circulation increased from large doses of aspirin at onset it improved recovery chances, less damage done to nerves and tissue. Acupressure relieving energy blockage in the head preventing major damage. Within three months he was able to resume the majority of his life. Normally that process takes six months to a year or more to recover. **Note:** tongue deviation occurs when a person is having a stroke. It points in the direction of the stroke. The tongue resolves to straight when the repair in the body becomes normal. Sometime I can catch a predisposition or a mini stroke just by looking at the tongue.

Stroke Case Study: A 71 year old male was driving along the highway and had to stop. His legs would not work. He was taken to the hospital. Signs and Symptoms: his left side was affected by a stroke. He had slurred speech, drooping left eye and lips, left hand hard to grip and to maneuver arm, left leg slightly affected in ability to lift it to walk straight. Diagnosis: Aortic Stenosis, Carotid Stenosis. He was on Metroprolol, Amlodipine, B12, aspirin 81 mg. His energy was okay, good appetite, his cognitive was slow and hard for him to respond to questioning. He was a smoker. He said he was having a hard time reading and processing any kind of input; verbally or visually. Handwriting was an issue because he is left handed. **Treatment:** He proceeded to do speech therapy, OT, acupuncture once a week. Within one week the leg recovered. Within two weeks handwriting was better, reading was better but not 100%. Speech comes and goes with fatigue. Left arm better, stronger grip. Left shoulder pain since CVA, some dizziness. One month later his smile was better, speech was coming along about 80%, reading better by 70-80% improved, energy better, no dizziness, handwriting 50-60% better, two things a day wear him out. At four months states he's 75% back, 90% speech, his shoulder got a steroid shot and feels good. Six months Acupuncture treatment one time a month maintaining well, face gets tired and droops, he feels 90% recovered, ITM herb Gecko/Gingko for brain, ITM Restorative for liver and heart, calming.

Analysis: Swallow points were used for speech impediment. Stroke can take some time to recover. Stroke is a major shock to the electrical body; it's like being hit by lightning. The circuits are fried and it takes time to rebuild the electrical pathways. The sooner you can get acupuncture after a stroke the better the results. When I was treating this gentleman another person came in with a stroke that had been in the hospital for six months recovering with no acupuncture. It took longer to wake up the pathways and the results were not as good. The speech improved and ambulation improved almost back to normal, but the arm never recovered its grip and began to hang after time with no use or exercise.

Traditional Chinese Medicine has scalp acupuncture for stroke if the client is willing. I have seen phenomenal results in China but poor compliance in the U.S.A.

Hepatitis

I've treated 25 people, 12 females, 13 males, average age 49 years old.

Hepatitis A is a bacterial infection of the liver. It is gotten by dirty hands handling food. People pick it up turn yellow and are sick for about a month as it runs its course.

Hepatitis B is a viral infection of the liver, blood born and can only be gotten through blood to blood carriers. This can be through dirty needles with drugs, transfusions, and really rough sex.

Hepatitis C is a chronic virus of the liver that lasts more than 6 months. It can create more serious health problems, cirrhosis of the liver or cancer. Average incubation is 45 days until symptoms appear. The symptoms are clay colored poop, dark urine, fever, jaundice-skin turns yellow and the eyes yellow, nausea/stomach pain, loss of appetite, joint pain. Symptoms can last 2-12 weeks.

In Chinese Medicine this is an obvious liver problem. Acupuncture is used to boost immune system, herbs to boost immune system and clean out virus.

This has been a serious issue in our culture. Blood transfusion screening didn't occur until after 1990. Transfusions before 1990 were possibly exposed to Hepatitis B and C. The treatment for chronic hepatitis was Interferon shots, $10,000 a shot, that made people really sick afterwards, not a high success rate for reversal of infections either. Interferon was harvested from healthy blood serum. It was a generalized antibody to viruses, not specific to Hepatitis B and C.

Hepatitis C Case Study: A 34 year old male found out 6 months ago that he had a diagnosis of Hepatitis C. Presented with some cognitive issues but happy personality. He had dryness of mouth, shortness of breath, drinks tea, cokes and juice. He had a good appetite with some diarrhea. He sleeps 12 hours and wakes up groggy. He drinks a lot of milk, likes sweets but watches them, low fat diet, walks a lot, smokes 1 pack a day. He is 245 lbs. and lost 20 lbs. He is on Lithium, Thorazine, Interferon 3 times a week for 3 months, milk thistle, Vitamin C, D, dandelion herb combination for liver. He has allergies to pollen in spring and fall. **Treatment:** Acupuncture, ITM herbs, homeopathy, NAET

Analysis: After 3 acupuncture treatments he was not having chills or reaction to the Interferon shots. He was trying to stop smoking, his triglyceride count was 1200. His liver and kidney enzymes were also up. Started him on ITM herb Buplerum 12 to help the liver with cholesterol and to relax and cleanse it. After 3 months his breathing was better. At 7 months of acupuncture therapy 1 time a week we put him on ITM herb Salvia Shu Wu to reduce cholesterol; he started exercising to reduce weight, down to 4 cigarettes a day. At 9 months we started rotating herbal formulas known to kick out Hepatitis C, ITM herbs Eclipta, Eupolyphagia, Lotus leaf, Salvia/Ligustrum, Zhu Dan. Started working on high blood pressure of 150/100 and anxiety used acupuncture and herbs. 2 years later liver spots on legs were almost gone, other primary care says "he looks the best he's seen him". He is not as moody and very polite. Triglycerides were reduced from 1200 to 430. It turns out that Lithium was driving the triglycerides levels up so a change of cognitive medication occurred which had its own challenges to stabilizing the aggressive behavior. He started counseling for his emotional behaviors. His thoughts were clearer and better communication skills. Did NAET's to correct malabsorption of foods and pollen allergies. In 1999 he was tested and told he was Hepatitis C FREE! He was checked 1 time a year for 4 years in a row and he stayed FREE of Hepatitis C. All the client had used was acupuncture and herbs to clear his Hepatitis C out, still smokes, still walked and fair to questionable diet. It took 4 years to clear out the virus with herbs and acupuncture 2 times a month.

Hips

The majority of my clientele have some form of back pain, about 60%. Low back pain can result from repetitive use. Kidney deficiency due to burn out of adrenals/stress, emotional trauma/drama, allergies, menses/menopausal issues, cystitis, prostate problems, incontinence, asthma, abdominal issues; constipation/diarrhea can occur also.

The low back can either relate to a hot issue, left kidney or right kidney for cold. If either of these kidneys are down there is a temperature correction that needs to be made. Night sweats, anxiety and heart palpations are common with left kidney down. Digestive, Lung problems are common with right kidney down. Knee problems relate to the kidneys also and most often if there is a low back issue the neck will have issues of compression or pain as well.

Restless Legs: is usually a mineral deficiency that can be corrected by taking a good liquid trace mineral (100 items) It can also occur as a side effect of statins. Occasionally there is a nerve compression from sciatic which can be relieved by acupuncture.

Sciatic: is always related to a compression of nerves in the low back. The pain radiates down the gallbladder meridian or bladder meridian (sciatic nerve is underneath both) to the knees and out the 4th or 5th toes. The energy in the body follows flow lines that need to open off the toes. If the shoes are too tight or a person is not able to release that energy off the toes, hammertoe results. Acupuncture is used for best results 6-12 treatments.

Hips and leg pain 44 treated, 30 female, 14 male, average age of 54 years old.

Acupuncture can help with the pain of hip replacement.

Bursitis Case Study: A 49 year old female presented with right hip bursitis radiating to knee then to ankle and foot, four years ago it started. Diagnosis was low back, nerve and muscular problems. Bowels irregular with diarrhea, constipation, no energy, restless legs at night. She presents hot with cold feet, dry eyes, and anxiety. She eats a fairly balanced diet, loves sugar but watches it, 2-3 diet sodas per day with 1 cup of coffee. Hysterectomy because of cysts, issues with diabetes, cholesterol, allergies to eggs, grapefruit and sulfa products. She is on Metformin, Simvastatin, Ompralezole, Paxil, Vitamin D, multi-vitamin, fish oil, iron, and gabapentin for restless legs. **Treatment:** Acupuncture, ITM herbs, Homeopathy

Analysis: Statins are known to cause muscle tightness/constriction as a side effect in some people. She stopped the medication and the restless legs disappeared. She had mercury, chemical toxins, parasites, and viruses stuck in her system. She started homeopathy to get some of those underlying problems out. She had Coxsackie virus that causes pain in hips and thighs. She took Chem X to get out of her body some of the toxins that create constriction in the muscles. She took ITM, Restorative herb to clear the heat causing dry eyes, anxiety, cysts on ovaries. The alternating constipation and diarrhea indicates a spleen Qi deficiency which means she doesn't have enough kidney energy to cook her food in the body. Long term effects of diet soda and too much sugar create heat and anxiety. Acupuncture was used to charge up the system and relieve the pain. After the three treatments her hip was better, no more diarrhea, no allergies. After five treatments the hip resolved, no pain radiation to knee and foot, bowels were good, energy a lot better, no restless leg.

Low Back-Sciatic Case Study: A 59 year old male presented with 10 years bulging disc, left side worse. Lumbar Spine L3/4, L4/5 degenerative disease. Right sciatic to hamstrings, weak back. He also has dry eyes, eats a balanced diet, and seasonal allergies. He takes heart medications for high blood pressure, medications for heartburn/bloat, fish oil, and vitamin D. **Treatment:** Acupuncture, ITM herbs

Analysis: Sciatic improved after two treatments. We used ITM, Xanthium for seasonal allergies. ITM, Mans Treasure for energy for the kidneys. Kidneys rule low back pain and when deficient can create the weakness which allows bulging discs to happen. Olive oil was suggested for more oil in the system for dry eyes.

After six treatments, sciatic was a lot better, he was doing exercises for support. After six months of two times a month he went to maintenance of one time a month for a year. This problem was the result of repetitive use, he was a landscaper.

Patterns for improvement with pain can include a steady, longer time between pain, getting longer until the client is good. After one week no problems, we then can stretch the treatment to ten days then to two weeks, if good at that point, can go to three weeks or sometimes people need a maintenance treatment, one time a month or sooner.

Therapeutically longer than one month does not keep maintenance. An immune cycle is every three weeks for the immune compromised.

Immune System

The physical body is a collection of layers of organs that work together in a community to house a spirit. This body has its own identity that a spirit lives in. It governs, feeds, cleanses, protects and even organizes itself; it has its own identity, but it needs spiritual energy to run it. The body is your house for your spiritual being. The body is the blood in Chinese Medicine, the chemistry, the physical house. The parts in the house, the Qi and blood work together to have a spiritual experience in a physical body. They cannot live without each other. This is the Yin and the Yang, the basis of our whole being.

The Immune system is our protective device to keep our personal identity intact. It keeps us cohesive in our own selves. It holds the separation between me and you. It continually asserts an identity every minute of the day based on your personal genetics.

The genetics can be as variable as the stars. Determination of strength of body are genetic durability, how hard wired you are, disease, nurture verses nature.

Emotions, environment, spiritual proclivity all influence the strength of the body. Some of the negative influences can be modified by good diet, clean environment, and good spiritual practices. The body promotes homeostasis. The body **always** wants to float to its highest potential. The spirit needs to give the body that chance, the problem is we are not given manuals when we are born for best practice, so we learn as we go and hopefully don't muck up the system irreversibly. Nature has given us a lot of fixes for most problems; we are in process of evolving. The old classification of the body's immune system is based on five immunoglobulins related to the five organs in Chinese Medicine. Immunoglobulins are the guardians of the organ system.

Immunoglobulins:	Chinese Classification:
Ig D-1st defense	LU/LI
Ig A-1st defense	ST/SP
Ig G-lymph/mucosal defense	HT/SI, TW/PC
Ig M-base immunity	KD/BL
Ig E-Allergy, Parasites	GB/LV

Ig A and Ig D mostly work at scavenging the digestive system to keep it cleared out of pathogens. Homeopathy stimulates this layer of antibodies to rally the system when the body has failed to recognize the offending pathogen.

Ig G accounts for 80% of the antibodies, can freely pass through the placenta wall, occurs in the lymphatic system and mucous barriers.

Ig E is present in high quantities in allergic individuals and also shows up with parasite infestations. Allergic people, long term, end up with hyperimmune systems that exhaust the adrenals of their steroids, depleting Kidney energy causing great fatigue leading to autoimmune disease.

Ig M is the largest molecule and represents the base immunity that informs the thymus to make specified immunoglobins (acquired immunity). This is the layer vaccines affect.

The human body when functioning well runs like a well-oiled machine. Vaccines are programs that are put into the computer to protect that machine from infection. Herein lies the problem our immune systems do not come online until we are 2 years old. The infant/toddler has residual immunity from the mother especially if she breast feeds. Injecting foreign material into an open system with no natural defense developed creates problems in some individuals!

The question becomes how many programs can one apply to an individual before their computer crashes from overload?

When this is done too young, the individual deals with a compromised immune system all their life. Allergies increase, autoimmune increases, fatigue increases because of overactive system with too many programs running. What is a necessary vaccine protocol? Many childhood diseases are obnoxious, but do not kill and immunity lasts a lifetime.

There are **stages of illness** in Chinese Medicine that can help identify a pattern. Each of these stages relates to a pair of meridians that are layers of protection to the body.

Tai Yang- (SI/BL) **Posterior Body:** fever, cold attack of wind or cold

Yang Ming- (ST/LI**) Anterior Body:** hot, profuse sweat, high fever, flushed face

Shao Yang- (TW/GB) **Lateral Body:** alter fever and chills, dry throat

Tai Yin-(LU/SP) **Abdominal:** abdomen full, bowel/digest issues, no thirst

Shao Yin- (HT/KD) aversion to cold, restlessness, no thirst

Jueh Yin- (PC/Liver) hot pain in chest, no appetite, cold limbs, diarrhea/vomiting

The first three patterns are related to excess. The outside factors affecting the body, cold, wind, summer heat, dry, damp and fire. The 4th and 5th patterns are related to deficiency, which is a long term insufficiency of energy, chemistry, internal starvation. The last level is basic functioning of organs and is nearly impossible to recover health from there. Understanding what level a client is at helps stage a recovery plan. This can help a person look at long term health goals. Recovering health is usually not a quick fix. By the time a person comes into my office, usually they are into the 4th and 5th stage of illness, not knowing the signs of the previous stages or able to correct those stages.

Tangles: are hard to get rid of. They usually involve three different categories of disorder for the body. This keeps a person trapped in the tangle.

1-Allergy, Parasites, Virus (Ig E) Allergies deplete the body of energy to keep moving, cook the food, staying focused, keep basic functions of the body working, breathing. When the energy is low or depleted parasites jump on to clear the debris and then the viruses attach because they like living in the debris. Parasites like living on toxins. When a person is allergic there is a lot of toxic debris, a protein glob produced from the immune system trying to protect itself. The problem is that the immune system is fighting itself when it is allergy intolerant leading to an autoimmune disease such as Lupus, Sjogren's, M.S, Chronic Fatigue, Fibromyalgia, Arthritis. Ig E is the immune systems component that goes after allergy and parasites. **Treatment:** Correct the allergy and detox parasites and viruses.

2-Candida, Parasites, Virus (Ig A, D) Candida is a chronic yeast issue that takes over the organs in the body. It is a collective organism that has roots that permeate into tissue. It forms a matrix throughout the body with its own chemistry. It likes acidic environments. It promotes parasitical infestations that like to live on the waste from the yeast, then the viruses hop on board. When the toxic load becomes too great the body begins to die. This tangle appears in most chronic fatigue, autoimmune, cancer and underlying allergies, bowel issues, skin conditions. Low thyroid can open the body up to this tangle. Ig A and D are the immune systems component that scavenge to keep the body free of yeast. If there is no energy in the body to fuel the immune system, this system fails. This is the end game for cancer. **Treatment:** Detox the candida through strict yeast free diet for at least 7 years. It takes 7 years to replace all the cells in the body. Eliminate parasites with a cleanse/herbs at least 1 time a year, except in the beginning more often because of the yeast; clear out viruses.

3-Kidneys, Diet, Brain In Chinese Medicine the brain is ruled by the kidneys. The left kidney rules the will and the right kidney rules the will to carry an idea or action to completion. The Chinese believe that one comes in with a given amount of kidney energy/Qi from the parent. If the parent is deficient the child has less to begin with. The old wives tale was "lose a tooth for each child". When a person runs out of Kidney Qi they die, the only way to replace it is with good nutrition. If the diet is poor or malabsorptive this energy is not replaced. I find most of my clients come in to my office when they hit this Kidney Qi wall. If we don't unravel the deep seated causes the client does not get better, they degenerate into a deeper level of disease. (see immune system) **Treatment:** Acupuncture, tonify the kidneys (charge them up), correct diet- "Eat right for your blood type" has good information in the book to correct diet at a core level. Stimulate the brain function with word games, physically/mental activities. Correct any environmental issues in the household or work environment.

Incontinence

21 people treated, 19 women, 2 male at an average age of 60 years old. Incontinence relates to low kidney function in Chinese Medicine. Tonify the kidneys is the fix with acupuncture and or herbs. Kidneys take longer to improve than any other organ and depending on compliance of the client can take years of incremental improvement. Incontinence is usually the end result of long term overuse of kidneys due to stress of some sort. It can be used as the indicator of improvement when working on kidney health. Incontinence can result with prostate surgery and improvement is possible.

Incontinence Case Study: An 83 year old female presents with incontinence, leakage last six months, retains fluid/legs swell she occasionally checks her weight to see how much fluid is retained. Diagnosed with Interstitial Cystitis, she's had flatulence for 3 years. She is allergic to shellfish and fish oil.

She has had her gallbladder and appendix out plus a partial hysterectomy due to endometriosis. She watches her diet, no milk products or sugar and few carbohydrates. She is taking Synthroid, Estradiol, Glucosamine/Chondroitin for her knees, Vitamins C and D with caltrate, Potassium Chloride, Ceclor/ Z pack/Bactrim rotation for cystitis. **Treatment:** Acupuncture, ITM herbs, Digestive Enzymes

Analysis: After five treatments of acupuncture her bladder was functioning properly, no incontinence. The rotation of antibiotics was keeping the bacteria killed in the bladder, but also killed the bowel bacteria causing gas/flatulence chronically. Suggestions of probiotics but also fermented foods for gut bacteria. Probiotics are usually only for milk products, lactobacillus, bacillus bifidus only grow with milk products present. She had worsening knee issues related to kidneys, preceding the last six months indicating weakness there.

Infection

In Chinese Medicine there are 6 layers or **stages of illness**. When someone comes in sick the most superficial layer has to be addressed first or an illness can be pushed deeper in the body's system. Infection is usually the most superficial layer. It can be a cold, flu, pneumonia, staph, strep, a bacterial problem. Antibiotics work quite well and fast for these bacterial infections. The overuse has created some problems. Now we are seeing more problems with viralized bacterium. In other words the bacteria has hooked up with the viruses and are creating deeper layered infections that people are not easily getting rid of out of their bodies. This is what I am seeing more and more. The only way I have found to get rid of these deep layered viruses is through homeopathy. This has to be staged if there are multiple viruses stuck in the body. If there are multiple viruses usually these people are in a body system crisis, for example autoimmune disease, severe allergies, fibromyalgia.

Chinese Medicine does have herbal remedies to clear the bacterial and viral issues out of the body. This includes cicada, scorpion, centipede and earthworms to clear heat, toxins and congestion. These take time to use and in our present society of quick fix no one takes the time. It takes 3-21 days to get an infection properly out of the body. The infection if not gotten out then proceeds into a deeper layer to be dealt with later.

Keep the body high functioning! Maintain excellent health through good habits, great diet, and exercise, drinking clean water, clean food, happy thoughts and happy spirit. Life is a gift with good health!

Staph Infection Case Study: A 1 year old male presented with staph infection on his penis 3 months ago. Has had infection 3 times and 4 times at the crown of his head. He is breast fed. He perspires after naps and sleeping at night, eats organic and basic foods. He is fussy with fatigue. He's had Hepatitis A and B, DPT, HiB vaccines, Varicella, Flu, MMR, Pneu, Rotavirus. **Treatment:** Homeopathy, NAET

Analysis: Kinesiology tested him for some food allergies and candida. Staph issue instigated by the DPT vaccine. We administered the homeopathy for DPT. He was doing a yeast free diet. They called 3 months later the staph infection had cleared completely off all body parts. NAET 1 month later for grains due to bumpy rash on legs.

Insomnia

Insomnia Case Study: A 36 year old male presented with "Don't Sleep" for 10 years, energy down, red eyes-sclera. He sweats occasionally at night, hard to get to sleep, his diet is fair, sweets 2-3 times a week, drinks 4 cans of mountain dew a day, no medications or vitamins. **Treatment:** Acupuncture, ITM herbs-Restorative, Gecko

Analysis: He quit the caffeine and slept better the second night home. We used acupuncture to clear the heat that keeps a person up at night, changed diet with reducing caffeine, sugar and alcohol in this case the caffeine that creates heat in the body. ITM Restorative herbs to help clear heat, build up left kidney for energy. ITM Gecko was added at 3rd treatment to get more energy in system. By the 4th treatment he was sleeping well and his energy was much better.

Insomnia Case Study: A 67 year old female presented with trouble getting to sleep, hot flashes, gaining weight, joint pain at times, sinus trouble, history of breast cancer 2 times and 33 times treated with radiation, she tends to be hot; she is going to weight watchers. She is on Lisinopril, Vitamin D, Multi-vitamin, Calcium and Letrozole for 5 years chemotherapy for cancer. **Treatment:** Acupuncture

Analysis: This woman was hot from chemotherapy and radiation therapy. This heat would keep her up at night. She worked at a hair salon and was exposed to a lot of chemicals, which also create heat in the body. Heat causes joint pain and hot flashes. It tells me her left kidney is not cooling the body. She did not want to do herbs. It took 12 treatments to manage the hot flashes due to still taking a chemo drug. Insomnia got better as the hot flashes improved.

Insomnia: Treated 187 people, 124 females, 63 males, average age 50 years old.

Insomnia is aggravated by too much heat in the system. This can be created by drugs, alcohol, caffeine, sugar, poor diet, too much work, pain, cancer/chemo, allergies, anxiety, arthritic pain, fibromyalgia, headaches, hyperthyroid, some autoimmune disease virus. Anemia-lack of oxygen to sleep calmly at night can keep a person awake.

Kidney Issues

33 treated, 22 females, 11 males at the average age of 49 years old.

I use the Kidney School model of Chinese Medicine on everyone that comes through my door. It is a constitutional school of thought and I apply it to everyone while I treat their acute issue. (Refer to the information on Kidney School) Kidneys are the bottom line of the body in Chinese Medicine. They control the heart whether it will flip out into a stroke, TIA's, High blood pressure; fix the kidney and it fixes the heart. The kidney is the energy that helps the body to cook the food, calm the liver/anxiety/anger, enlighten the brain, builds the bones, controls asthma attacks, and grows hair on the head. It relates to low back pain, knee pain, osteoarthritis, heel spurs, and bottom of the feet pain, prostate issues, ED issues, menses, menopause, allergies, and fatigue and urination issues. Acupuncture and herbs bring up the kidney. This organ takes the longest to reboot of the 5 organs. Acupuncture can speed up the healing of broken bones and cut the process time in half.

Kidney Case Study: A 50 year old male presented with stage 4 kidney failure. His GFR (glomerular filtration rate) was to 10%, last 3 years under 20%, creatinine high. If he eats wrong, he can feel it in his kidney. Left kidney has no output, Right kidney working hard. He is on a waiting list for a kidney. He has had kidney stones in the past, has psoriasis, both knees hurt. He has a cold body and sweats at night, his energy can get really low and he naps when fatigued. He has a good appetite; eats low potassium foods, no beef, few veggies, some potatoes, some berries, watches his sugar, no alcohol, has sips of caffeine. He is a farmer. **Treatment:** Acupuncture, ITM herbs-Gecko, Kochia

Analysis: After 3 weeks he was tested and his GFR had gone up a point. He was excited! His knees were better after

3 treatments, after 4 treatments the psoriasis started to recede and the tingling in the feet was gone. After 9 treatments no neuropathy, knees pretty good-sore with work, energy better depending on how hard work was that day. The Kidneys were holding their own.

Note: GFR is the test for how well the kidney is processing urine. It indicates the healthiness of the kidneys.

Kidney Failure Case Study: A 69 year old male presented with kidney failure, colitis due to antibiotics, lost 20-25 lbs., cough, post nasal drip, not able to urinate-building up in body-can't let it go, on dialysis 3 times a week, does physical therapy 3 times a week. He has a very cold body can't get warm, shortness of breath, thoracentesis 3 times a week for fluids on the chest, A-fib palps at times, insomnia, eating a very alkaline diet, has little meat, sugar, no alcohol or caffeine. Has a history of 3 open heart surgeries, appendix, partial thyroidectomy, 3 thoracotomies. On Crestor for 30 years, Levothyroxine, Coumadin, Doxycycline, Clindamycin, B-complex, Probiotics **Treatment:** Acupuncture, ITM herbs Restorative, Man's Treasure

Analysis: After 2 treatments he was not as fatigued, less frequent bowels, nausea better and less of a cough. After 3 treatments he was able to urinate better, less nausea and vomiting, his sleep was better. At 4 treatments he had noticeable energy, sleep improving, bowel stabilizing, little shortness of breath, and urination somewhat better. He took herbs for both kidneys; at that point he left to go to Mexico to get warm.

Knee Pain

Treated 55 people, 34 female, 21 male, average age 58 years old.

Knee pain is related to kidneys in Chinese Medicine. Tonify the kidneys with acupuncture and herbs. One should see improvement within 3 treatments; if no improvement it is a longer term issue and will take at least 6-12 treatments. Knee pain is more superficial in the hierarchy of maladies for the kidney. Acupuncture on the knees relaxes the muscle attached ligaments so there is less bone on bone rubbing. Chondroitin and Glucosamine help to rebuild the collagen on the bone surface so it moves more smoothly. Rheumatoid knees are a dietary allergy issue, corrected with NAET, no more arthritic build up and acupuncture can gently detox the crystals built up in the joints. Acupuncture can help with pain after knee replacements. The elderly tend to have sore knees from kidney deficiency. One can have bone spurs in the knee to stabilize it.

Knee Pain Case Study: A 62 year old female presented with knee pain due to overuse 10 years ago. The knee had been scoped and rooster comb collagen shots applied. She hurt worse with walking, change of weather and prolonged use. She was on statins. **Treatment:** She had acupuncture 5 times with electrical stimulation to relieve the pain. Her energy was improved. She took ITM Restorative herb 1 time per day.

Analysis: She was in pretty good shape after 5 treatments and did not return. Several years later she did get a knee replacement.

Knee pain can occur because of overuse/over work, which seemed to be the case here. When evaluating a client other options may be in the mix. Knees in Chinese Medicine relate to the kidneys. If the kidney energy is decreased one can have knee pain. Knees also relate to the low back pain. This can be determined by examining the pulse to see if the kidneys are down. Sometimes low back treatments fix the knees or vice versa. Sometimes straight energy treatments fix the knees and or the low back. Statins can also create leg and knee pain.

Knee Replacement Case Study: A 64 year old female presented with left knee replaced, tight thigh into hip (1 year ago), right shoulder collarbone broken and needs more range of motion, right knee had cortisone shot-helped, constipation, low energy, tends to be cold, watches her diet, history of disc fusion lumbar 4/5. Vitamins include omega 3, EPA and DHA. **Treatment:** Acupuncture, ITM herb Restorative.

Analysis: After 3 acupuncture treatments her range of motion in her shoulder was greatly improved and then wears off before next treatment. After 4 months the shoulder remained good with no further treatment. Her knee improved within 2 treatments and then the pain went up and down with intensity with more use, resolving after 3 months. The more issues, the longer it takes to resolve both of them together. The body will heal what is easiest to fix first and it may not be what the practitioner keeps asking the body to do. She continued to do maintenance 1 time a month for the next 6 months. The ITM Restorative herb was for low energy and the constipation.

Lungs

In Chinese Medicine the stomach is the 1st line of defense, not the lungs. The Stomach Qi/energy breaks down the food to strengthen the body from the inside so a person doesn't get sick. The lungs and large intestines are filters for nutrients for the body. Lungs for oxygen, large intestine for water and miscellaneous nutrients missed by the small intestine. These 2 filters keep a healthy body working. Sometimes the filters need to be cleansed; colon cleanse, massage, soaking, ocean beach time, exercise with a lot of water. When the filter gets clogged due to allergy, impure air, poor diet, bronchitis and sinus issues occur. Long term congestion brings down the immune system. Fatigue is usually a result of not getting enough oxygen in the body. The lungs are the easiest to correct in Chinese Medicine.

309 cases of bronchitis, 233 were woman, 76 men. Most of these had several rounds of antibiotics. Acupuncture can boost the immune system so the body can fight off the invader. Herbs can strengthen the immune system preventatively or help move the allergen, toxin or bug out of the system. The acupuncture and herbs can also help calm the histamine system from overreacting as in asthma or anaphylaxis, allergies. Homeopathy can help rally a tired immune system. NAETS can reduce the body's histamine reactions to allergens, not only pollen, outside allergens, but chemicals and food allergens. A hyperimmune system can overreact to allergens and cause a flood of white blood cells to the lungs creating so much mucus a person can't breathe.

Bronchitis Case Study: A 60 year old male presented with chronic bronchitis, asthma to dust and pollen, sinusitis with cold weather and allergies. He had a history of pneumonia, arthritis, and pleurisy. He tends to be cold, shortness of breath, coughs with hacky phlegm. Past was a smoker for 20 years, but stopped for 29 years now. He has insomnia and had ear tubes for drainage. He is on Advair, Singular, Multi-vitamin, Omega 3, Vitamin D, and Boswellia. **Treatment:** Acupuncture, ITM herbs Pinellia, Perilla Seed, Gecko, Restorative, Siler, Kava, Kroeger homeopathy, NAET

Analysis: He had several viruses stuck in his system. 60-92 is a strep virus in the lungs. I usually find this in every chronic bronchitis case. The immune system goes down with trauma, colds or in this case allergies and then this opportunistic virus embeds and any time there is an immune hit it goes after the lungs. We first started the homeopathic to get the 60-92 out then added herbs for energy (Gecko). He used kava for sleep. We started NAET to peel off the allergies. We used acupuncture to boost the immune system. Siler was used for swelling legs, Forsythia for sinus until we got his system up enough. Pinellia was used for the thick heavy phlegm and cough. Perilla seed for asthma and Restorative to energize and help with insomnia. The herbs for the lungs were staged depending what was going on with his body. It took 4 months every week to stabilize and then every 2-3 weeks for 2 years. He still comes in for maintenance every once in a while.

Bronchitis Case Study: A 26 year old female presented with chronic bronchitis 6 times a year. In childhood she had colds a lot with tubes in her ears, she was on a lot of antibiotics. She's allergic to milk products and pollens; she eats a lot of sugar, no fruit or veggies. She's not rested and blurry eyes looking at a computer all the time at work. She has acne from a poor diet. **Treatment:** Kroger homeopathy, ITM herbs.

 Analysis: She was sick for 6 weeks on antibiotics with chronic bronchitis. We gave her 60-92 strep virus in the lungs to start. She also had lymph flukes congesting the lymphatic system, and coccidiomycosis which is a form of valley fever (mold) she had a DPT (whooping cough vaccine) residual from her shots (Pertussis vaccine for cough) and 2 other viruses in her system. She lived 2 hours away so we had to stage the homeopathic, and straightened out her poor diet. After 6 months of homeopathy and diet change we put her on herbs for kidney support and herbs for lung support. Her energy improved. It turns out her house was full of mold. She moved and improved rapidly.

Bronchitis Case Study: A 69 year old female presents with chronic bronchitis every in year in spring. She had a history of mono and fibromyalgia. Signs and symptoms: insomnia, night sweats, and all over pain. She had 4 viruses stuck in her system, parasites in lymph and mercury in the tailbone. Mono initially opened the gate for multiple viruses to enter her system. She is a massage therapist and has been exposed to a variety of illnesses over the years. Parasites are a yearly cleanse that needs to be addressed in general, causes buildup of congestion in the lymph and lungs. Mercury in the tailbone comes from mercury in the tooth amalgams. Some people have leaking fillings and after 50 years of old fillings there may be leakage. Mercury in the bones, tissues can cause fibromyalgia anywhere in the body. The viruses can also cause fibromyalgia and the chronic bronchitis. **Treatment:** Acupuncture, ITM herbs, Kroeger homeopathic

Analysis: The viruses are taken out homeopathically 1 at a time. This takes approximately 1 month to do 3 drops 2 times a day. The homeopathic rallies the immune system's white blood cells to go after the offending virus. When testing the client, when they reach a 40-60% level of immune activation there can be a herxheimer reaction or healing crisis where the white blood cells go after the virus and get it out of the system. She cleared the viruses out and has not had bronchitis or fibromyalgia for 6 years.

Memory

Chemical Toxicity Case Study: A 77 year old female presented with indigestion, heartburn, depression, anger, emotional, can binge on sweets. She had a radical hysterectomy from fibroids and endometriosis. She had three children and two miscarriages. She has seasonal allergies. She lives on a farm with some chemical toxicity and viral issues. **Treatment:** Acupuncture, homeopathy

Analysis: In Chinese Medicine kidneys rule the memory. An analysis of her history shows that she has heat, dry eyes, depression, anger. Sweets become hot in the body's system. Fibroids and endometriosis are heat in the uterus. Stress can cause the heat, too much sugar and or emotional turmoil. Heat affects the liver which needs to be calmed by the kidneys. Long term chronic issues will drain the kidney energy which is the vital force of the brain. Underlying kidney issues will have an alternating diarrhea/constipation problem. Migraines are heat rising to the head, which can be from stress or food allergies. Children and miscarriages all take energy away from the kidneys. Lowered immunity can occur from toxic exposure and open the door for viruses to jump on. For treatment she started out every two weeks and has maintained that cycle. When she misses or takes a break she regresses; mentally getting more spacey and forgetful, Chinese Medicine calls that "cotton head".

Menopause Case Study: A 50 year old female presents with forgetfulness, sinus problems, digestive issues, insomnia, menopausal aches from waist down, and wiped out. She tends to be hot, dry eyes, hot sweats, and short of breath. She has had surgery on both knees. Her sinus drains all the time and she is on estrogen replacement **Treatment:** Acupuncture, ITM herbs, Homeopathy, NAET

Analysis: We began with acupuncture to boost a depleted system. She had four viruses and chemical toxicity. We started homeopathy one vial per month per virus if she could handle it. Her energy got better. Kidneys rule the "will" and the "will to carry it out", basically how much energy you have. Knee pain is related to kidneys as the knees improve one can tell the kidneys are improving. Restorative herbs will decrease the heat in the body-hot sweats, dry eyes, some digestive issues, such as heartburn. The Restorative herbs works on the left kidney, which is heat/hot flashes, and helps menopause. She quit aching in the abdomen, forgetfulness improved as we took off the load of too much heat in the system and charged her battery through acupuncture. Food allergy improved with NAET so she could get her nutrition to help the brain function properly. When someone has sinus all year round this indicates food allergy. First look at milk products, then grains and yeast. It took about six months.

Alzheimer's Case Study: A 74 year old female presented with diagnosis of Alzheimer's disease. Her care takers requested an evaluation for her moaning and screaming with pain. She had progressive Alzheimer's over 12 years. She was cold with cold extremities. She had stomach pain. She had been diagnosed with diverticulitis in the past. She had night sweats for the past 3 months, possibly a sinus headache. Her affect appeared blank with occasional focus. She drinks a lot of water with no edema, has constipation, sleeps about 14 hours plus a nap. She had 5 children and a hysterectomy. She had no teeth. She favored eggs and sweets-cakes, breads, pudding and included meat, veggies and fruit. She was exercised daily for range of motion. She took Mellaril, a psychotic drug, Dilantin for seizures, Flexeril for muscle spasms, Nubain for bad spasms and Tylenol with codeine for pain. **Treatment:** Acupuncture, ITM herbs

Analysis: ITM herb, Zhu Dan given to clear out liver/gallbladder pain, any sludge in the liver. Acupuncture helped her immediately no pain for 30 hours. She was able to burp. She would have spells of pain and sometimes the shot/opioid didn't work. Her pain got worse after her gallbladder surgery 3 years ago. After the 4 treatments she was able to move her bowels every day. Acidophilus, enema implant were given to restore bowel bacteria. She became more alert. We food tested her for 130 foods to see which ones energetically supported her. After 2 months, bowels started to move by themselves and she was sleeping good. After 6 months of 1 treatment a week she would start laughing and singing. She became more alert when her husband came over. The untold story was after 12 years the husband was living with another woman an R.N. and this client would get extremely agitated when the women (R.N.) came over. They eventually sent the client away to a home and she outlived the husband.

Lung congestion and Large Intestine were her primary issues for maintenance therapy once the digestion issues were cleaned up with good diet, smaller portions, compatible food.

Menses- Menopause

32 treated for menopause average age was 48 years old. Quite often women would come in with an issue menses oriented with their other primary medical concerns. Going through menopause can cause minor physical problems to become larger problems such as migraines, autoimmune, emotional issues, hot flashes, constipation, thyroid issues, arthritis, palpitations, low energy, herpes, anemia, fibro breasts, increased allergies, insomnia.

Menstrual pain can occur at any age. The female reproductive organs are ruled by the liver and the kidney. The kidney builds the blood the liver moves it. Stress, emotional turmoil and poor diet can restrict the liver's ability to move the menses. This constriction results in pain because the blood is not moved out. Compression of restricted movement long term causes endometriosis, polycystic ovaries, fibroids, pelvic pain. When the body starts to slow down towards menopause the constriction still applies but the body is hormonally confused because of lack of hormones to push a cycle that occurred for many years.

Diet is key in supporting hormones. Poor diet as a youth leads to anemia, leading to no blood production for the liver to push therefore pain.

In Menopause the adrenals take over for the ovaries production of estrogen and need the help of a good diet to support the adrenals. Chronic stress and emotional turmoil suppressed create fibroids, endometriosis and breast nodules. Chronic yeast infections are related to a poor diet and or a partner's poor diet. Yes, a woman can get yeast infections from their partners, so can men. (see candida)

Hot flashes can occur at any age, but mostly at end stage because of the left kidney not having enough juice to get through menopause. Women can get stuck here and seek hormone replacement therapy or get acupuncture and herbs to correct the problem or end up getting a hysterectomy.

Pelvic Pain Case Study: A 14 year old female presented with stabbing pain right side spread across whole abdomen, spiked a fever, eyes dilated, liver enzymes elevated, shortness of breath, low energy, cold with hot flashes last few nights, little appetite, depressed, craves sugar. History of regular menses flows 5 days, 2 days before cycle pain and then 1st day pain and pea size clots, she doubles over and vomits. She was put on BCP's lost 15 lbs., she lost her appetite, became more tired and cold. All vaccines were current; Hep B, MMR, no Gardasil.

The previous week was in the hospital for extreme excruciating pain. There was no diagnosis given and she was released. She was given Levaquin, Flagyl, Doxycycline, 800 mg Motrin, Morphine, Fentanyl, and Zofran. She presented weak and pale. Mother's history indicated polycystic ovaries and ablation. **Treatment:** Acupuncture, ITM herbs, Women's Treasure.

Analysis: The birth control pills provoked the ovaries into a cystic explosive state with extreme pain manifesting. She was taken off the birth control. One year later all is well with no pain during menses, appetite returned, no shortness of breath and participating in normal activities. Good diet needs to be maintained for long term health of kidneys.

Menopause Case Study: A 41 year old female presents with menses, bleeding a lot/clots, spotty last 3 months. Signs and symptoms she's hot, has shortness of breath, fatigue, insomnia, and sad.

She was examined extensively with western medicine and couldn't find an issue for excess spotty bleeding. **Treatment:** Put on ITM Imperata a Chinese Medicine to "stop bleed" and she was good within 2 months, no bleeding. Did acupuncture and ITM herbs.

Analysis: This woman appeared to be going through menopause at an early age. Her shortness of breath, fatigue and insomnia was related to anemia due to excess bleeding.

Shortness of breath and sadness is related to decreased lung function related to Kidney Qi deficiency. She did acupuncture 1 time a week and moved from the area. She had called and said she was okay within 3 months.

Menopause Case Study: A 40 year old female presented with menopause, hot flashes since 17 years old, headaches 1 time a month, pain, diagnosed with GERD, was intubated 3 times scoped for digestive issues, polyp removed but created esophageal constriction. Gall bladder, appendix, bladder polyps and colon polyps removed. She had leukemia when she was 3 years old. Allergies to Bactrim and "clothing stores"-formaldehyde, she is hot, has blurry red dry eyes, tinnitus with left ear ache, fatigue, belching and gas, emotionally overwhelmed, and eats small meals so she doesn't choke. She likes sweets but watches input, has little milk products, and has 2 cups of coffee a day. She takes papaya, digest ease, fish oil, birth control, and citalopram for anxiety. **Treatment:** Acupuncture, ITM herbs, homeopathy, NAET.

Analysis: After 3 acupuncture treatments and ITM Restorative herbs the hot flashes were a lot better. She cut back on her coffee to clear the heat. Her throat constriction relaxed, she was also using castor oil on her throat to reduce inflammation and swelling. She took homeopathy for viruses, mercury and chemicals stuck in her tissues creating fatigue and pain. She took ITM Women's Treasure herb for energy and various ITM herbal combinations for sinus and cough. We treated a bunion 2 times and pain went away and the bunion got smaller. We did NAET for heartburn/GERD digestive pain, belching and gas. After 1 year she feels pretty good. She had gout symptoms with her legs swelling and bloating, after she did NAET caffeine the signs and symptoms went away for that issue.

NAET-Nambudripad's Allergy Elimination Techniques

NAET: Reprograms the body not to be allergic to anything known or unknown. It can also help with fibromyalgia, rebuild tooth enamel, osteoporosis, motion sickness, heat, cold, intolerance and lack of absorption issues. NAET (see process under NAET evaluations)

230 plus people treated, 176 Females, 54 Males average age of 44 years old.

I have treated in utero children to 93 years olds to improve basic lifestyle. Allergies can affect many parts of the body. The greatest issue is intolerance and malabsorption of foods, chemicals and outside irritants. This malabsorption affects the functioning of the organs therefore **dis ease**.

The autoimmune can be allergic to their own organs. A person can be allergic to heat, cold, humidity, air-conditioning, low or high altitude, clouds before a storm, dry heat, exercise, motion sickness. There is combination NAET's that can help different problems. Vitamin C + blood vessels for Raynaud's Disease, amino acids + stomach acids for hyperactivity/autism, sugar and brain tissue for brain fog fatigue.

A collapsed immune system activates all sorts of allergies. It can aggravate viral and parasite issues. Tangles suppress an immune system. One has to start somewhere to climb out of the lower stages of illness.

NAET can clear out and help repair viruses from Lyme's Disease, overdose from an anesthesia reaction, hormonal issues, arthritis, esophageal burning, asthma, heart palpitations, sugar addiction, weight gain, migraines, electromagnetic fields, fluorescent lights, computers, all skin problems/acne, bowels, sinus, lungs.

Type I diabetes Case Study: A 5 year old female presented with Diabetes type 1 on an insulin pump, checked for celiac and was negative, teeth trouble rotting easily. She is hot and has night sweats. She has had tubes in her ears for allergies; she has nightmares, and eats a regular diet. She had RSV (respiratory syncytial virus) as a toddler; vaccines are all up to date. She takes Lantus overnight and Novolog in the day for her type 1 diabetes. **Treatment:** NAET, homeopathy

Analysis: She had 3 viruses in her system a shingles, mono and retro virus. She had a toxic load of dioxin, pesticides and preservatives. We cleared these out with homeopathies. She struggled with controlling her blood sugar. Viruses and bacteria infections are known for driving the blood sugar up. She began to balance out better after a year of NAET with her blood sugar. Her blood sugar improved, she was fragile. Her father was a farmer and was exposed to a lot of chemicals. Toxic chemicals can bring down a child's immune system to pick up viruses. Mother worked in health care.

Diet is a key factor in a fragile diabetic's world. They need neutral foods that don't push the blood sugar to extremes. Meat, root veggies (a lot of vegetables), few carbohydrates, little sugar, no caffeine or alcohol. Occasionally a little sweets is okay.

Stomach Pain Case Study: A 14 year old female presents with stomach hurting all the time since birth, sometimes sharp, stabbing pain, flat affect and went to counseling and it helped. Diagnosed with Dyslexia, Dyscalculia, she likes boundaries; she has safety issues. She likes to be warm, watches diet for food antagonists. She takes Zofran, probiotics, uses essential oils, heating pads for pain. Had allergic reaction to HPV vaccine and had mono 3 times. She had tubes in her ears as a child and a stomach biopsy. **Treatment:** NAET, Homeopathy, Acupuncture, ITM herbs

Analysis: She had a parasite issue and several viruses in her system that we got out with homeopathy. Herbs were used for sinus conditions until we got through the NAETs. After 3 months she was doing well and off all medications.

The underlying viruses can keep a pain syndrome going. Allergies to food and pollen were indicated by having tubes as a child. She emotionally improved with less stress from pain in her stomach. Emotionally from a Chinese Medicine perspective, she couldn't "Digest" her own life.

Allergies Case Study: A 32 year old female presented with frontal headaches and migraines. She was hot day and at night, had cold sores, can't breathe well, swelling ankles and no energy. She watches her diet and craves salt, she has trouble with belching and gas, has seasonal allergies. She takes fish oil and prenatal vitamins. She's had her tonsils/adenoids, and gallbladder out. **Treatment:** Acupuncture, ITM herbs, Homeopathy

Analysis: Acupuncture helped with the fatigue, hot flashes and swelling ankles. She did NAET for the allergies and homeopathy for viruses. ITM herbs were used to help with her energy and eliminate parasites. She found out she was pregnant in the middle of NAETs; of her 3 children when the baby was born it had very few allergies versus her other 2 who had a lot of allergies.

Neck and Shoulders

Neck and Shoulders: treated 128 people, 72 females, 56 males at an average of 54 years old.

Acupuncture can loosen up a frozen shoulder, improve range of motion, reduce neck pain sensations from amputation and neuropathy, reduce neck pain from arthritis, whiplash, stenosis, concussion, and stress. Neck pain can be directly related to low back pain because of the curvature of the spine. Headaches can result from underlying neck and shoulder constriction (thoracic outlet syndrome). Spurs in the neck for stability can cause pain. TMJ can come from neck pain.

Acupuncture can help chiropractic adjustments stay in longer. Acupuncture can help with nausea due to tight neck/migraines.

Neck-Stenosis Case Study: A 45 year old male presented with right shoulder tear, right elbow pain, neck stenosis, low back compressed discs and radiates down legs. He complains of chronic headaches. He takes Aleve for headaches and anti-acids; wakes up with pain at night. Diet is balanced, drinks a pot of coffee every day. **Treatment:** Acupuncture-electro stimulation

Analysis: After 4 acupuncture treatments for the neck, it was a lot better. The lower back radiated a little and the arm needed more. After 2 treatments the bunion had no pain. After 6 treatments no radiation of low back pain and it was just a little tight. The neck was ok, shoulder still needed more work. After 14 treatments the shoulder was doing well. 6 years later he returned with low back bulging disc, neck tight, right shoulder had surgery on it. Bunions both needed help. He was a professional roper.

Pain Case Study: A 51 year old female presented with degenerative disc disease. She had neck surgery in 2001 for C4/5 and C5/6. The vertebrae were plated and screwed together. Now C 3/4 is bulging with a fragment pushing out on the right side. This pain radiates down the right shoulder; pain from the bra strap to the neck. She has a high pain tolerance but it is now raising her blood pressure to 140/98, when normal was 90/60. She tends to be hot, dry eyes, has migraines 3 times a year. Lyrica helped with some of the pain which is an arthritis drug. She cannot tolerate opioid drugs, therefore she tried acupuncture. **Treatment:** Acupuncture and ITM herbs

Analysis: Pain caused in the neck by compression of vertebrae tends to follow the meridian pathways of the body.

In this case it follows the Triple Warmer and Bladder Meridians.

Heat in the system can often cause more pain in the upper body. Her history indicates long term heat in the system, dry eyes and migraines. Diet can cool heat. She presented a diet of a lot of caffeine and sugar which creates long term heat in the system. Normal would be 1 dose of caffeine per day from any source, for example coffee or soda; green tea is ok in profuse amounts because of the pycnogenol contents. A lot of milk products can also lead to congestion in the system that creates heat. 1 dose of sugar and alcohol a day is okay.

Osteoporosis

Osteoporosis: 35 treated, 33 females, 2 males, average age of 65 years old. Osteoporosis is a calcium deficiency in the bones that make them weak and more brittle as a person ages. There are many factors in this process. The end result is a broken bone that is hard to recover because if one bone is weak, where can a broken bone get enough calcium to repair itself from the stores within the body? The kidneys tell the bones to absorb calcium. The body needs calcium, vitamin D, iron, trace minerals to make bones strong. If a person kidney's are weak and not communicating to the bones there is an issue. If there is no absorption of nutrients the bone can't be built. The bones are ruled by the kidneys in Chinese Medicine.

Older people as a rule cannot process their food as well as they did when younger because of poor diet, bad habits, malabsorption or intolerance of food, food allergies.

Side effects because of shortage of calcium in the body are muscle tightness/twinges, fibromyalgia, blood clotting, maintaining a normal heart beat, fatigue, insomnia, parathyroid issues, no absorption can see excess calcium stored in joints (arthritis or kidney and gallbladder stones)

Osteoporosis Case Study: A 75 year old female presented with arthritis in fingers, left hip pain/sometimes it radiates down leg, straight neck radiates to shoulder, fatigue, she is cold and has had both knees replaced. Her diet is balanced, she has 1 diet coke a day and she exercises 30 minutes 2 times a day. She is on Lisinopril, Simvastatin, Amoxican and Meloxicam and Tylenol for pain. She also takes a Cal Mag D, Multi Vitamin, Glucosamine, Omega 3, and aspirin 81 mg a day. **Treatment:** Acupuncture, ITM herbs, NAET

Analysis: Acupuncture was used for arthritic pain in neck, hip and fingers. She was able to sleep longer without hip pain with 1 treatment. She stopped her statin and felt better without it, memory sharper with less muscle pain. She had no pain in fingers after 3 treatments. She was put on ITM Restorative herb to improve her energy, and support kidneys. She stopped her diet coke and calcium supplement. She started ITM Siler herbs for knee swelling. After 5 months of treatments she was feeling good. Her cholesterol started going up and she was diagnosed with Osteoporosis. We began NAETs. When she was finished after 4 months, her cholesterol came into normalcy and her bone density test came back normal. She had corrected Osteoporosis. NAET reprograms the body to absorb correctly the nutrients it needs and let go of unnecessary products.

Acupuncture can heal broken bones faster. I had a compound ankle/leg broken bone come in who played soccer and wanted to get out on the field sooner; we did acupuncture on the leg that didn't have a cast and he recovered within 3 months instead of a diagnosed 6 months.

Acupuncture will stimulate bone growth for any joint replacement, cuts healing time in half. One still needs physical therapy and occupational therapy for long term recovery.

NOTE: I have seen people get better from arthritis of the fingers when they stopped their calcium supplement. The calcium supplement was too much and the body started depositing the extra in the knuckles and other joint structures forming spurs.

Pain

Acupuncture is primarily used to reduce pain, most of my clients come in with some sort of pain. A course of treatment in general is 12 treatments for 1 issue. I tell people to come in for 3 treatments and if it works…then YEAH! If not, then we don't waste each other's time. Usually the 1st treatment can go 3 ways: 1. No change, 2. Feel better for a day or two then goes back to old pattern or 3. Feels worse, then better then goes back to old pattern. After 3 treatments we will know if the problem is an acute one or chronic. Chronic takes more treatments. We keep applying acupuncture treatments until a person is good for a week and then stretch it out to 2 weeks. We can then go for 3 weeks (an immune cycle). Maintenance is usually every 2-3 weeks with a chronic pain issue or repetitive use. Therapeutic herbs can help with pain syndrome also.

It's important to find the constitutional issue underlying the problem, sometimes it is not just mechanical. It can be a virus, poor diet, bad habits, energy too low, kidney deficiency which isn't getting the endorphins to the brain or natural steroids to the body to neutralize pain naturally.

Sometimes a primary issue is resolved and then another underlying issue appears. The body wants to be well! If a person is in severe pain, treatment everyday can occur, for example tooth pain until one gets to the dentist.

Pregnancy

The primary goal of Chinese Medicine for fertility is to boost the kidneys. The kidneys rule the womb, the hormones to regulate the womb and the energy to produce a child. It can take anywhere from 6 months to 2 years to improve the kidneys enough to conceive. This depends upon the mother's ability to recover the kidneys and maintain the Kidney Qi while pregnant. Multiple miscarriages indicate that the kidneys are too weak to hold a child in the womb. I have worked with women to get in condition to conceive and hold onto the child. Acupuncture and herbs build a person constitutionally to bear children. I have worked with at least 15 women who produced children.

Acupuncture can be used to induce labor as well and naturally. It can be used for morning sickness with great success. It can be used to keep the woman's energy up to prevent a miscarriage.

Menses Hormones are regulated by the kidneys. The right kidney works the Yang/progesterone and the left kidney works the Yin/estrogen. From the menses (Yin) to ovulation the hormones are regulated by estrogen, from ovulation to menses (Yang) the menses are regulated by progesterone. Where ever the trouble is in proper menstruation is the kidney that needs to be supported. If periods are too short in between or too long ITM herb, Nuphar is great to regulate that. Normal is approximately 28 days with no cramps or clots when menstruating. It usually takes 3 months to rework the menses. Diet is a factor correcting the menses. Too much heat creates endometriosis, fibroids, pain and clots. Too much cold creates pain, vacant feeling and no conception.

Infertility Case Study: A 33 year old female presented with allergies and fertility issues. She is a farmer and has issues with hay fever and mold. She can get right sided migraines, anxiety, constipation, face and hand puffiness and wants to quit smoking. She has no energy tends to run hot with cold hands and feet, her eyes get red, dry and blurry with allergies, she gets anxious and shortness of breath with irritable stress, sweats and can't focus. She eats a lot of processed food and primarily carbohydrates. She feels like she's getting a head cold 2 days before menses and gets puffy. She has bled for a month or two now after the last miscarriage. She's been taking fertility drugs and nothing has worked. She takes prenatal, probiotics, B complex and Vitamin D. **Treatment:** Acupuncture, ITM herbs, homeopathy, NAET

Analysis: We put her on ITM herb, Women's Treasure to build blood, and invigorate the kidney. She's also taking ITM herb Nuphar to regulate the menses. She tested for 2 viruses and chemicals in her system; we then started her on homeopathy, 1 vial a month to slowly reboot the immune system. Acupuncture was 1 time a week to get her immune (energy) system up. Did some NAET's for her to better absorb her food. Used ITM herb Acorus for anxiety. It was 2 years since her original visit she became pregnant and made it to full term.

Most women with infertility issues have a very depressed immune system. The hormones are not being regulated. This is a kidney issue. Too much stress, smoking, alcohol, sugar and caffeine burn up the kidneys. Dietary change needs to occur. Building blood with eating meat, spinach, root vegetables and blood foods. It takes time to regenerate the body. Acupuncture maintenance can help keep the energy up in the body while pregnant so there is less likely a miscarriage. Herbs can also be used to keep the kidneys up.

Infertility IVF Case Study: A 28 year old female wants to conceive, she presented with hashimotos, hypothyroidism, Vitamin D deficiency, high blood pressure, ADHD, irregular menses, constipation, low energy, she tends to be warm, chronic eye dryness, insomnia, balanced diet, 1 cup of coffee a day, allergies to pollens, popcorn and gluten. She takes Levothyroxine, Prozac, and Strattera. **Treatment:** Acupuncture, ITM herbs, homeopathy, NAET

Analysis: She is put on ITM herbs for her kidneys, ITM Restorative and herbs for regulation of menses, ITM Nuphar. She has parasites, chemical toxicity, mercury, radiation and virus issues. We start homeopathy to slowly remove these issues 1 per month. She had been bitten by a brown recluse. We worked on the thyroid to boost it and the kidneys for energy with acupuncture. It took a year just trying to get her immune system up. Her menses stabilized. Used ITM herb Restorative to clear the heat. She tried IVF once and it failed, she lost the fertilized eggs. We began NAET's to improve the chemistry in her body. After 3 years of treatment she was able to conceive IVF and retained eggs to completion.

Hormones Case Study: A 41 year old female presented with adrenal fatigue/exhaustion, sciatic, diagnosed with chronic fatigue syndrome, fibromyalgia, high Epstein Barr virus count, mono as a child, allergies to pollen, molds, and milk. She wants to conceive. She has a cold body, hands and feet, dry red eyes, insomnia, eats a balanced diet. She has cramping and clotting after menses. She's had 2 miscarriages. **Treatment:** Acupuncture, ITM herbs

Analysis: She used primarily her own herbal combinations. We did acupuncture for 2 years and she was able to conceive and bring forth a child. We warmed the abdomen and moved the stagnant Liver Qi that was creating a disharmony between her cold body and hot liver. (Dry red eyes and insomnia)

Morning Sickness Case Study: A 27 year old female presented with vomiting bile, she's 32 ½ weeks pregnant, has pressure in digestive area, carrying high, no energy, morning sickness with the first two children. She's had her gallbladder taken out, has seasonal allergies, borderline depression, craves sugar, balanced diet, no alcohol, has 1-2 cups caffeine, she tend to be hot. **Treatment:** Acupuncture

Analysis: After 1 treatment she was able to eat, no nausea, some heartburn remained, her energy was better after 3 treatments we stretched to every other week until birth.

Prostate-PSA-ED

Enlarged prostate and the PSA test are the defining factors of malfunction for men. A high PSA is considered a cancer situation in men. Chinese Medicine does not look at it that way. They use acupuncture and herbs to bring down that high PSA value. ITM herbs Lindera and Chihko Curcima have been known to bring a high PSA value into normalcy. In men, the prostate and women the breasts are soft tissue that the body can store toxicity into. They are both on the liver meridian, so if the liver is too toxic it deposits into softer tissue. The process of elimination of toxicity starts at the liver, then the kidneys and last of all the skin. The soft tissues accumulate as much as they can and then they signal the body with lumps of encapsulated toxins. The viruses like to assimilate in these. The tissue is encapsulated, a form of basic protection of toxicity from the body. The immune system can't get to it unless it's cut open "Surround the Dragon: If you can't get to the lumps; one uses herbs to erode the shell and go after the irritation. Herbs take about 3 months for complete results. Acupuncture can speed the whole process up.

ED (Erectile Dysfunction) ED happens when the kidneys wear out. This can be from poor diet, bad habits, malabsorption of nutrients, or just hitting the "Kidney Wall" as we age. Acupuncture and herbs can build and recharge the kidneys. It can take a few months or years to recharge and once the wall is hit it needs to be managed. The Chinese have a schedule for optimum orgasm and age. There are Qi gung and yoga practices to support the kidneys in older age.

PTSD

10 treated, 4 Female, 6 Male, average age of 50 years old. Acupuncture works well for anxiety, TBI, relaxation of body, mind, spirit. Also, counseling can help move issues on. Worked with PTSD around murder, stalking, and Vietnam War.

Allergies Case Study: A 54 year old male presented with allergies, joint pain, acid reflux, body is warm, dry eyes, shortness of breath with exertion, insomnia, balanced diet, drinks a lot of coffee. He's taking Prilosec and Allegra for a year. **Treatment:** Acupuncture and ITM herbs

Analysis: After 3 treatments the acid reflux went away unless he had a stressful week. After 3 months he had no joint pain, slept well, no shortness of breath, his eyes were ok, cleared the heat in system and no allergies. He went to maintenance every other week, then to every 3 weeks. He was the first to the scene of a suicide of someone he knew. His joint pain and acid reflux returned after the traumatic experience. We used acupuncture to relax the body and clear the mind. He went to counseling and 2 months later he said he was doing much better.

NOTE: Each emotion corresponds to an organ and vice versa. His kidneys (fear) and his heart (anxiety) were treated. Counseling is always important to deprogram the issue out of the body.

Seizures-Epilepsy

20 Treated, 13 Females, 7 Males, average age of 36 Seizures and Epilepsy are related to the kidneys. The brain is fueled by the kidney energy. Tonify the kidneys and the brain works better. There are classic herbal formulas to enhance

the brain, ITM herb Bu Nao Wan, Gingko good for circulation of blood to and through the brain. Stress and poor nutrition can cloud the brain; the Chinese call it "Cotton head"

Acupuncture can help to realign the batteries-Kidneys. NAET can help with malabsorption of nutrients that aren't being absorbed.

Epilepsy Case Study: An 18 year old male presented with a history of 104° fever and in bed for a week and then sick for 3 weeks at 10 years old, he then proceeded to have a petit mal epileptic seizure. His digestion is gassy and he is short on energy. His eyes roll up when he goes into seizure, the medicine stops the eye rolling but he still has the seizure. He had an EEG to confirm Epilepsy diagnosis. He is told he can't have a driver's license unless he is on Depakote. He occasionally takes Vitamin C and a Mega Vitamin. His diet is balanced but eats little fruit or vegetable and eating schedule is erratic. He works at a high-end restaurant as a specialty cook-fast paced. He enjoys outdoor activities. **Treatment:** Acupuncture and ITM herbs.

Analysis: After 4 acupuncture treatments his energy was better. ITM herbs, Restorative to balance the brain. I put him on ITM herb, Omphalia for parasites because further investigation showed he had been in India and had diarrhea on return for months. Further history indicates he had fevers at 12, 13 of 100-102° on and off medication for Epilepsy and the incidents would happen when he was tired. He was not good in the morning, sometimes with nausea. He needs at least 8 hours of sleep. He gets bronchitis occasionally. We food tested him to 130 foods to see what foods supported him energetically and what foods to stay away from that drained his battery.

Epilepsy seizures are related to the kidney in Chinese Medicine. People who experience periodical seizure have a system malfunction that shocks the kidney battery and drains it. These people need a lot of sleep to recharge. Kidney time is the weakest in the morning from 5-9 am. A person should rest, read, meditate, and do quiet time to recharge. Kidneys are strongest 5-9 pm at night and a person usually perks up then. A weak kidney person needs to eat a lot of protein, veggies, and some fruit and carbohydrates, little sugar, alcohol and caffeine. They need to eat often and regularly with small meals to keep the engine going all day. Rest heals the kidneys if they have been overused. Stomach is most active between 7-9 am and if overactive can be nauseous or hypoglycemic if not enough food has been maintained in the body's system. He was stable for 4 years.

Seizures Case Study: An 11 year old female presented with history of seizures during the summer. She has mid-back pain, pollen allergies, her diet consists of a lot of sweets, pizza, toaster strudels, meat, likes salty and fried foods. She likes to exercise. Her ears buzz and she can have shortness of breath with allergies. She drinks water but is constipated, she sleeps well and has not started menses yet. **Treatment:** Acupuncture, ITM herbs, NAET

Analysis: Diet change was first to address to support kidneys. (See previous Epilepsy) ITM herb Restorative to charge up the kidneys and brain, also to cool the liver heat that was giving her mid-back pain. Restorative can also help with ear buzzing (Kidney's with Liver component) and constipation. She was food tested to 130 foods for best energetic foods for her and energetically draining foods. We did NAET's for a few basic foods. She had not started her menses, so hormonally a seizure could have come on if there was too much drain on the Kidney; for example over exercise, poor diet/habits, not enough rest or sleep. She maintained on herbs and when older after 12 years old received acupuncture to stabilize the body and spirit. If the body isn't maintained properly the spirit can't sustain itself in the body.

After 4 years she had learned to monitor herself. She watches for pupil dilation that equals weak kidneys. She tends to have seizures waking up so she tries to get enough rest. She has tested the limits of her body with over exercise, caffeine, sugar, oral surgery, alcohol. Too much of any of these things aggravates seizure activity.

Skin Conditions

Treated 61 cases, 41 females, 20 males at an average age of 48 years old.

There are 3 layers of internal cleansing. The liver is the first layer to clean toxins out of the body. Kidneys are the second layer and skin is the last layer and largest organ of the body. When some form of rash appears on the body, cleansing needs to occur internally. To correct **Eczema** or Psoriasis the internal constitutional cause needs to be found and corrected if possible. **Psoriasis** is known for "heat in the blood". Allergies, chemical toxins, internal heat, stress, virus load, hormonal imbalance, damp heat, immune system down, candida, spider bites, toxic building exposure, deodorant, emotional trauma all contribute to congestion of the skin. **Rosacea** occurs on the cheeks when allergies, especially chemical allergies are present. **Hives** occur with any kind of allergen and need to be taken seriously because you can have them on the inside of the body and choke to death. Recovery takes time and needs to be staged.

Eczema Case Study: A 25 year old female presented with a right medial finger joint that had raised bumps, dry and scabby. Diagnosed with Eczema, she has a cold body, occasional tinnitus, balanced diet, craves sugar and eats it, she's allergic to latex. **Treatment:** Acupuncture, ITM herbs Picrorrhiza

Analysis: She used castor oil topically to reduce inflammation. Olive oil was suggested for lotion after a shower. Yeast free diet was suggested. Rash on the middle finger relates to the Triple Warmer in Chinese Medicine. This rules the lymph system. If the lymph system is congested, it's usually a yeast overgrowth. She just had a baby and her hormones could be still reorganizing which can cause a yeast issue. She called 3 days after the first acupuncture treatment and was going through withdrawal signs and symptoms from yeast. She had changed her diet, increased her protein and the next day had better energy. It took 4 months on yeast free diet for the eczema to go away, by then her energy was good, sleeping good, doing well, just tired of the diet.

Eczema Case Study: A 60 year old male presented with a serious skin reaction to weeds. Big water blisters on hands, back and legs. He was given steroids and creams to fix it. He never completely cleared up. He was exposed to chemicals as well, that he then threw away. This farmer is exposed to various products that go on the fields. He has no energy. He presents with neuropathy in his feet after cancer therapy, (testicular) chemotherapy for 3 months. His eczema was around his knees, back, lower legs, ankles, top of feet, spots at belly, elbows, hands, typically at the joints. I suggested castor oil to be applied on all areas affected, it reduces the inflammation and swelling. **Treatment:** Acupuncture every week for 2 months and then every other week thereafter until resolved. Maintenance every 3 weeks when able. He was put on ITM, Restorative to charge the system up and ITM, Kochia to reduce itching and dissipate eczema. After 1 month the spots on his back were almost gone, spots on ankles receding and neuropathy on feet were better. After 2 months his skin was 80% better. He plateaued at this point so we changed the herbs to ITM Plantago, these herbs work specifically with rashes on the legs. With this combination and ITM Restorative he was able to completely heal all the eczema and improve the foot neuropathy to 1-2 incidences every 3 weeks. He maintained a schedule of every 3 weeks for a year to get through the worst times of calving and farming. New chemicals for the fields aggravated his issues but with greater care in handling the chemicals and increased acupuncture he was able to deal with them and the hints of skin issues receded.

Analysis: Stress always increases the signs and symptoms. Farmers/Ranchers deal with a lot of stress especially during certain times of the year. Acupuncture can help calm the body when overstressed so the body can function properly. Acupuncture also boosts the immune system and helps resolve depletion in this case the previous health issue in dealing with cancer.

Rash Case Study: A 55 year old female presented with rash of stomach progressing to left leg. She has left hip pain, ruptured L4/5 10 years ago, heartburn, insomnia, and allergy. Rash is itchy and worse at night. She is hot with warm hands and feet, with occasional night sweats, shortness of breath with too much exercise, she eats a lot of popcorn, licorice, coffee, and diet coke. She's allergic to sulfa and penicillin. She takes a multi vitamin, tums, Aleve for pain, potassium and steroid cream for rash. **Treatment:** Acupuncture, ITM herbs, homeopathy, NAET

Analysis: Suggested castor oil for rash, and no diet artificial sweeteners. She took ITM herb Women's Treasure for energy. She did homeopathy for viruses in system 1 a month. Acupuncture helped with the hip and had no pain after 4 treatments. Started NAET for rash and after 16 allergy treatments rash went away and she no longer had heartburn. Acupuncture cleared the heat stuck in the body, she was able to sleep at night and was not as hot, and she also corrected her diet.

Spider Bites

Spider bites are treated with homeopathy and acupuncture. Acupuncture is used to keep the immune system up and circulation moving. Homeopathy is used to boost the white blood counts to get the specific pathogens out. This works for mosquito and tick bites as well. (West Nile, Lyme's Disease, Rocky Mountain Fever) Treated 8 people, 6 female, 2 male, average ages were 47 years old. The viruses that spiders carry can cause flesh eating deterioration, internal infection, chemical sensitivity, push a person into autoimmune disease, blisters, and burning-neuropathy.

Spider Bite Case Study: A 68 year old female presents with diagnose of M.S from 15 years ago, signs and symptoms of degenerative discs, osteoporosis, right leg drop-drags it, headaches, heartburn, retains fluid/swelling, bronchitis 1 time a year, cold, body pain, she states she has had "46 surgeries", allergic to a lot of medications. **Treatment:** After 6 treatments her headaches improved, sinus resolved and musculoskeletal system got better. She was put on ITM herb Alisma for fluid retention and she was able to walk further without pain and continues to use the herbs to this day; greatly reducing swelling.

Analysis: The first treatment session she attended a small black spider crawled out of her pant leg and onto the table. At first I was horrified that there was an insect on the table, I did not entertain bugs in the office and had never had a problem with them before. The disturbing part was that every time she came there remained a small black spider on the table after she left.

My disturbing conclusion was that these spiders may be the cause of her M.S if her house was infested. M.S is a neurological disease and can be caused by virus, toxic environment, spider bites?

Stones

Kidney, Bladder and Gallbladder stones, all of these are treated with ITM herb, Lysmachia, which breaks down the stones, softens them to be flushed out of the body. I had one person use this with pancreas stones with success. I had another gal who would come every 2 years to get a bottle of

100 tabs, take the whole bottle and it would soften her gallbladder stones so she did not have right side pain (where the gallbladder is). If the stones are too large, greater intervention needs to occur. This herbal combination is good for small stones and sludge. Apple juice with pectin, 1 quart a day for a week can help cleanse the system. Ultimately a better diet corrects the issue more permanently. Less milk products and acid foods, more vegetables. Most people don't eat enough vegetables!

Stop Smoking

Stop Smoking Case Study: (look under protocol under addictions)
A 32 year old female presented with sinus infection, low energy, and occasional migraines with hormones at menses time, insomnia, diet of meat, carbohydrates, cheese, caffeine-coffee 2-3 times a day, regular exercise. She's been smoking since 18 years old and wants to quit. **Treatment:** Acupuncture protocol NADA in ears and body points to support immune system. ITM herb Ardisia to detox and Echinacea for sinus infection.

Analysis: When treating "Stop Smoking" I ask for them to come every 3 days in a row and then spread the treatment out to 1 time a week. If a person hasn't cut back in 1 week to quit, the odds are it's not going to happen. This gal was perfect every 3 days for 3 treatments, then stretched for 1 week and then for 2 weeks. She continued to take the Ardisia and was 35 days clean the last time I saw her.

Teeth

Teeth: Treated 12 with teeth or gum issues. 10 female, 2 male, average age 38 years old.

Teeth in Chinese Medicine relate to the kidneys building bone. Gums relate to the stomach meridian. I have many persons with tooth pain come in to correct that pain with acupuncture, whether it is before a procedure or after. I have been able to save teeth that were hit hard in an accident that were loose. I had 2 people with abscessed teeth that healed up with "oil pulling" an Ayurvedic application. These teeth were good 2 years later. Through NAET, enamel was able to grow on adult teeth of children who had tooth enamel problems. Corrections of allergy to dental work happened with NAET. Bad teeth can show up with ear and eye infections, trigeminal neuralgia, tooth grinding is related to stress and can be helped with regular acupuncture to destress the system, and treat the anxiety.

Teeth Case Study: A 33 year old female presented with tooth pain from a student who had head butted her in class, knocking her over. Her neck was sore, whiplash and her right front and right lateral teeth were loose and in pain. She had lost weight from not being able to eat. She was emotionally lethargic and had TMJ with head pressure and range of motion issues. **Treatment:** Acupuncture, homeopathy, Hypericum.

Analysis: Acupuncture was used to release the pain and heat in local area at teeth. Body points to calm the body system. She got physical therapy exercises for the whiplash and acupuncture as well. Her whiplash range of motion improved, no complaints after 4 Acupuncture treatments. Her front right tooth had become grayish. She was still sensitive to cold after 5 treatments but without pain. After 8 treatments her teeth were feeling good and color had returned to the tooth, there were no zingers. She came in 1 time a month for maintenance for 6 months. A year later she returned due to some sensation in the tooth for 2 treatments. Then 1-1/2 years later for 3 treatments to support the tooth again for some sensitivity. There was never any infection just nerve trauma.

Tooth Enamel Case Study: A 9 year old female presented with inability to mineralize teeth. She was given all her vaccines between 3-5 years of age. She had no known allergies, no herbal supplements. **Treatment:** Acupressure, NAET, homeopathy

 Analysis: We used NAET to reprogram the body to absorb calcium and minerals to form teeth. She had a strep virus, fungus, mono and cryptomyrosis issues that were treated homeopathically. She had been sick with strep throat all winter long. We did NAET for 15 items. Her teeth remineralized. Her sister had the same issue and repeated treatment for her with great success.

Thyroid

Thyroid rules the metabolic system. It tells your body how fast or slow to cook the food. It's a regulator gland. It relates to stomach/spleen in Chinese Medicine. In a toxic world thyroid is the first contaminated. It's located in front of the throat and exposed to radiation first. Mullein can be used as a tea tincture to clean it out; Vitamin C stimulates the thyroid, silica aides in thyroid function. Sodium Fluoride takes the iodine out of it. Essential oil Myrrh is used for a hyperthyroid and Myrtle for hypothyroid. Don't fast if your thyroid has an issue. The biggest sign and symptoms of thyroid are weight gain with hypothyroid and severe weight loss with hyperthyroid. Hashimoto's is an autoimmune issue with the thyroid; a chronic lymph infection in the thyroid (attacking itself). The thyroid eventually wears out with long term stress, allergy, poor diet, and chronic yeast infection.

Hyperthyroid Case Study: A 52 year old male presented with mono, thyroid ultrasound nodule large, joint pain, low energy, asthma with exercise and cold, heartburn with acidic foods, insomnia, body tends to be warm, cold hands and feet due to frostbite, tinnitus last 2-3 years, balanced diet, has coffee 3-4 times a day, likes sugar and has it least 2 times a day. He takes Albuterol/Advair as needed for asthma, multi vitamin, omega fish oils, primrose, vitamin C and herbal reflux vitamin combination. **Treatment:** Acupuncture, ITM herbs, homeopathy

Analysis: Started homeopathic for mono first to get virus out of system and ITM herb, Restorative. Thyroid issues always have an adrenal feedback component when there is an issue. The pulse tells me what kidney is down and we start supporting it. In this case it was diagnosed as hyperthyroid and he didn't want to do radiation, he wanted to try the herbs first. ITM herb, Prunella to stop hyper functioning thyroiditis and ITM herb, Laminaria to cleanse the lymphatic immune system. A diet of root vegetables and meat, suggested more salt to support kidneys and check progress with the iodine test. (See evaluations) He lived 2 hours away so herbs were primary option. He was a professional in a small community under a lot of stress and worked long hours. Within 2 months he was feeling better. After 1 year the diagnosis of hyperthyroid was changed to a non-issue.

Thyroid-Hashimoto Case Study: A 41 year old female presented with hypothyroid 15 years ago, diagnosed with Hashimoto's, inflammation of the joints, interstitial cystitis 10 years ago, Irritable bowel syndrome diagnosis 20 years ago, she has headaches every 2 weeks, low energy, and wants weight loss. Her body tends to be hot with cold feet. Headache at base of skull every 2 weeks, constipation, menses normal, has 4 children, balanced diet, craves sugar but watches it. She is on Levothyroxine, Metformin, fish oil, vitamin D, B-12, and Progesterone. She watches certain foods for the cystitis, avoids alcohol, chocolate, onions, citrus and tomatoes. **Treatment:** Acupuncture, ITM herb, homeopathy

Analysis: She was put on ITM herb, Restorative to boost kidneys, clear heat, to clear headaches, get more energy, and fix constipation, due to heat. Homeopathy was used to get rid of chemicals-dioxin, pesticides, preservatives and viruses that cause joint inflammation, IBS, depress the thyroid and can create cystitis. After 5 Acupuncture treatments, homeopathy and herbs, she had no complaints of headaches, her energy was better, brain less foggy, less joint pain. We switched to NAET to work on the food allergy component for cystitis, IBS, chronic joint inflammation, thyroid issues. After a year of NAET some acupuncture and herbs, completing the homeopathy she said she was very well. She had lost weight, no headaches, no joint pain, bowels working well, good energy. She moved into a new house with her 4 beautiful children and happy ever since!

Thyroid-Hyperthyroid Case Study: A 63 year old female presents with arthritic pain in neck, carpet tunnel in hands, stress, low back pain, feet and ankle swelling, stress, she runs hot with rash on wrists and top of feet. She's bloated and had a rapid heartbeat. She was diagnosed with Grave's Disease with no nodules.

After 2 months of acupuncture her digestion settled down, she had no bloat or gas. Heart medication was regulated by a M.D., medication given for hyperthyroid. We began doing thyroid Acupuncture points which helped regulate the metabolism. Swelling decreased, swallowing improved, and bowels began to work better. It took 4 months to get thyroid normal range with acupuncture every week. Her energy improved, she had several viruses stuck in the system and used homeopathy to remove Epstein Barr, DPT, radiation and chemicals (Dioxin Pesticides Preservatives) She went to acupuncture every 2 weeks after going 1 time a week for 4 months. It took 2 years for the thyroid to go into remission and off all medications. She maintained acupuncture every other week due to stressful night work. **Treatment:** Acupuncture, ITM herbs, Homeopathy.

Analysis: She had other problems that were treated concurrent with the thyroid issue. The acupuncture improved low back pain, carpel tunnel and neck pain. Hyperthyroid issues can create heat-hot flashes and rashes due to heat. Digestive issues of bloating, diarrhea and constipation are all related to thyroid. Underneath at a constitutional level is kidney deficiency or energy to cook the food move the bowels and causes bloating. Arthritis can also be an issue because the foods aren't being digested and the residue from the food is going to the joints.

TMJ

TMJ: Treated 32 clients, 30 Females, 2 Males, average age 45 years old.

TMJ can be caused from head trauma, headaches, TBI, concussion, whiplash, tooth pain, tight neck and shoulders, stress, grinding teeth, dental procedures, braces and can cause ear ringing.

TMJ/Whiplash Case Study: A 43 year old female presents with ear and jaw tightness into neck from whiplash, hard to sleep, she has a warm body, hands and feet, with hot sweats at night. She's had migraines in the past, eye redness, ringing ears that are sensitive to cold wind. Her diet is balanced; she craves sugar but watches it. She is worried and irritated emotionally. She had mono in high school. Her cholesterol runs high 325; she is on a statin and multi vitamin. She is a book keeper. She has had all her childhood vaccines, a history of 10 days with a kidney infection as a child and UTI's. Her menses is normal. **Treatment:** Acupuncture, ITM herbs Restorative.

Analysis: After 2 treatments her TMJ resolved and her sleep was good. She had heat in her system that needed to be moved out of her body to correct the insomnia and hot sweats. The herbs and acupuncture helped with ear ringing and red eyes. Yoga and meditation was suggested to destressing from her work. Her diet was normal for such high cholesterol – dietary suggestions and ITM herbs to bring down cholesterol values were suggested. Further investigation into cholesterol issue would be worth figuring out.

UTI-Urinary Tract Infection

UTI's: Treated 10 clients, 9 Females, 1 Male, Average age of 47 years old.

Chronic UTI's can be caused by Kidney/Bladder stones, infection, too acidic body system, poor diet, food and chemical allergies, not enough water to flush the body, IBS, diverticulitis, chronic yeast infections, sexual partners, hysterectomies creating empty spaces and the bladder falls, diet soda, too much sugar caffeine or alcohol, STD's.

Kidneys rule the bladder and their infections. The kidneys need to be boosted with Acupuncture and herbs.

Chronic UTI Case Study: A 38 year old female presented with vaginal/bacterial infection, itchy, hurts with urination, continuous throbbing on fire, first in urethra, 5 rounds of antibiotics. She was diagnosed with a positive Cocci infection. She had a sore throat on and off. Her body is cold, gets headaches when neck is out of alignment, drinks water, and watches her diet "Weight Watchers" balanced diet. She has 4 children all by C-sections, seasonal allergies. Taking Relive herbal products and probiotics vaginally. **Treatment:** Acupuncture, ITM herbs, Homeopathy

Analysis: Started homeopathy for strep virus for her and her family, 2 children were carriers. She and her husband had a herpes and retrovirus combination; we started them one at a time after the strep virus was cleared. She was put on ITM herb, Restorative and Women's Treasure to boost the kidneys. After 6 acupuncture treatments the bladder was good. A year later she was back with some irritation and was put on ITM herb, Pyrosia for long term interstitial cystitis. She likes keeping up with maintenance 1 time a month. She gets worse with stress at harvest time.

Vaccines

A community consensus determines what the greater good is for a group of individuals for mutual protection in their environment. This is artificial immunity created by vaccination. It is an outside source dictating to the body an artificial barrier. The body has to work harder to maintain this vaccination program by constantly reproducing the program through the Thymus, white blood count complex. It has to keep running the program through the lymphatic/blood system, always alert always on guard. The kidneys are the organ that informs the blood system to reproduce white blood cells. The Triple Warmer/Pericardium/Lymphatic support the Thymus for the programing. The Spleen cleans the lymphatic system. The liver provides the nutrients, base material and enzymes to produce the white blood cells.

Rules of Engagement:
1: Be aware of how many vaccines dosage and frequency you are willing to give your child or yourself. Often the children are given adult dosages. Remember the child's immune system doesn't come on line until 2 years of age.
2: Is the vaccine necessary? Hepatitis B is for a virus that can only be transmitted through blood, for example blood transfusions, dirty needles-drug addiction or really rough sex. Hepatitis A is a virus that is found only in food from dirty hands handling it. A person is sick for a month, but it is not deadly or deforming.
3: Don't give a vaccine if a person is sick or feeling unwell. The immune system is already engaged in battle, don't add another antagonist.

4: Don't vaccinate if there is an autoimmune issue. This just gives more material for the body to attack its own system.

Most of the childhood vaccines chicken pox, measles, mumps, rubella, molluscum, hoof and mouth are obnoxious viruses that build natural immunity, if the child gets the disease. There are some complications but rare. The autism rate has increased, from the 1960's 4 in 10,000; in 2014 it had increased by 15% over the 2012 total, in 2018 1 out of every 60. (**CDC**) Center of Disease Control (**U.S. government report of Autism**) What has changed culturally? GMO came in the mid 1990's, MMR vaccine in the 1970's. Is it some other factor? Is it a combination of factors?

When multiple vaccines are given in a short time the body doesn't have time to assimilate. Remember it takes 3 weeks for an immune cycle. That means making the antibodies to protect self from 1 virus. An overabundance of foreign material can crash a fragile system. How strong is the immune system at engagement? Don't treat if sick.

Flu vaccines are chosen each year by retrieving 3-4 viruses from China to inoculate our population, sometimes the right ones are gotten, sometimes not. This is still another immune **program** put into a person's **computer.** Is the body strong enough to accommodate another program?

Some people have really strong constitutions and the vaccines don't bother them. Some people are more sensitive to these vaccines and need less. Elderly persons are more immune compromised.

If one keeps their immune system strong much of this is not necessary. (See Immune System Chapter)

The tribe that does all the vaccines is creating a viral reservoir or soup. This reservoir has snippets of all the previous vaccines floating around in their systems. When mutated within a body this can create new disease. (Bubonic Plague) "Black Death"? What are we setting ourselves up for?

A person legally has a right to conscientiously object. Your body is your own. Parents oversee the bodies of their children, they are the housekeepers until the child is old enough to care for themselves. Be aware and make good choices. Don't allow bullying for your choices.

Vertigo

Vertigo: 6 treated 4 Females, 2 Males, average age of 54 years old. Vertigo is an underlying issue for a lot of different problems. Chinese Medicine says it's a kidney issue. Balance can be a blood pressure problem, anemia-lack of blood to pump in the system, as simple as needing to drink more water to push around the body or as complicated as the otoliths in the cochlea of the ear need to be reset because the cochlea dictates balance to the body in a spacial environment. Trauma and age are the biggest contributors to vertigo issues. Stroke concussion, whiplash, Bell's palsy, anemia can cause vertigo.

Vertigo/Anemia Case Study: A 43 year old female presented with Vertigo. 3 years ago, a sinus infection originally started it, a month ago she was out for 3 days nauseous, couldn't function, physically sick, foggy brain and then exhausted for days afterward. Her energy is down and she is anemic. She is cold with cold hands and feet. She has heart palpitations 3-4 times a month and she is anxious. She drinks a lot of water and eats a balanced diet with 2 cups of coffee per day, she craves sugar. She takes Allegra for pollen allergies, a daily vitamin, Vitrex for Vertigo and Fluoxetine for anxiety. Treatment: Acupuncture, ITM herbs
  Analysis: The underlying anemia creates the anxiety and heart palpitations, not enough blood flowing through the body to get enough oxygen to the brain and the heart. Long term infection can make a person anemic because the body has to catch up and replace the dead red and white blood cells. Sinus infection can affect the balance in the ears. The brain is trying to reorient balance which creates exhaustion after so

115

many days of reorienting. Lack of blood can create cold in the body. She received ITM herbs Restorative and Women's Treasure to build kidneys and blood. After 4 acupuncture treatments there was no vertigo, her energy was better less foggy brain. After 6 treatments she reported 100% back to normal.

Vertigo/Allergies Case Study: A 61 year old female presented with vertigo, stomach issues-upset and too much acid, weak bladder and low energy. She was hot body and feet, shortness of breath with stairs. She got constipated with gluten, eats meat, little vegetable, cheese, craves sugar but watches it, loves salt, active on the farm. She had a total hysterectomy due to endometriosis. She has a cholesterol issues. She is on Yoli, low dose cholesterol medication, thyroid medication. She has known food allergies and allergic to alfalfa. **Treatment:** Acupuncture, ITM herbs, NAET

Analysis: She was put on ITM herbs Women's Treasure and Restorative to warm up the digestive system and give her more energy. It helped her move her bowels better. We started the NAET process immediately because her underlying vertigo issues were malabsorption of nutrients to create blood, therefore anemic. Stomach upset and acid went away because she was able to digest the foods. She corrected her diet to add more veggies less milk products, reducing the cholesterol problem. Cholesterol issues can also be malabsorption of fats, when that is corrected then the fats are absorbed and used where the body needs them. A crave for salt is trying to correct the kidney. After 16 treatments she was feeling a lot better, her food was absorbing, she could eat less and feel satisfied. No complaints of vertigo.

Viruses: Cold Sores, Herpes, Molluscum, Mono and Shingles

In the 1960's bacterial infections were more common than virus. Antibiotics were developed to combat the bacterial infections with great success. Antibiotics were developed in the 1940's. Overuse of antibiotics has pushed bacteria to mutate into viral entities. We have an epidemic of viral diseases now that antibiotics won't touch. We have a couple of anti-viral medications that suppress the immune system to tap down viral outbreaks-for cold sores and genital herpes. Mononucleosis is the gateway virus that tends to depress an immune system so more viruses jump on. Chronic Fatigue to Autoimmune Disease occurs over the span of a lifetime. I have had as many as 6 viruses on a person plus a tangle of chemical-pesticides, dioxin, preservatives, mercury and then parasites all making a perfect environment for virus leading to cancer. Virus likes damp/heat in the body. It's important to get rid of one or the other and preferably both.

A strong immune system will not pick up the viruses. Good, clean diet, toxic chemical free environment, good choices around vaccines, exercise, healthy spiritual life, all contribute to excellent well-being.

Mono: 140 treated 109 Females, 31 Males, an average age of 46 years old. (Read Immune System-Tangles, Vaccines)

Mono Case Study: A 16 year old female presented with weight loss 25-30 lbs., protein in the urine, lost her appetite and menses. She was diagnosed with mono. She was put on prednisone for laryngeal spasms and asthma. She was bloated with foods from steroids. Her body was cold with cold hands and feet, she slept a lot and had trouble with constipation. She was very athletic before the mono. She eats a balanced diet and watches the sugar. She was given the

Gardasil vaccine twice and thought the mono could have come from a weak immune system after the vaccine. She had all other childhood vaccines. She takes Vitamins and Essiac tea to clean out the system. **Treatment:** Acupuncture, ITM herbs, NAET, homeopathy

Analysis: Homeopathy was started right away to get multiple viruses out; a Nerve virus, Epstein Barr virus, DPT, Gardasil. When kinesiology is used, one tests which virus is causing the specific body discomfort. She was okay with the other vaccines, although I have had virus issues in other people come up later in life, after we cleared the 1st layer. Some of these viruses you can get again when the body system is down. All of the family might carry the virus but only certain persons show signs. Check the family! She was put on ITM herb Women's Treasure to energize and support the digestive system. The herbs can also warm the body up. She did 16 treatments for NAET allergies so she could start absorbing her food again. Viruses tend to screw up the digestive system so it can't break down the food and absorb it. Protein in the urine indicated a kidney dysfunction, she has no kidney battery or it's working poorly. We added ITM herb Tang-Kuei tabs to help the menses and Rhubarb for the bowels. After 6 months of treatment she had good energy, back to school participating in sports, eating happily, bowels normal, still waiting for menses to restart. It takes time for the food to get assimilated in the body so every part works right. It takes 7 years for all the cells in the body to be replaced.

Herpes: are cold sores and genital herpes. These are viruses stuck in the body that reoccur with stress once a person has it. Treated 21 clients, 15 Females, 6 Males, average age of 46 years old. Stress, poor diet, poor habits tends to aggravate herpes into erupting. Menses/hormones can precipitate herpes.

Usually the 1st time getting sick with the virus a person feels really bad and later with reoccurrence the virus doesn't bother as much but the body can't seem to permanently kick it out.

Homeopathy rallies the white blood count to get the virus completely out. It takes about a month. The practitioner can check percentage rates to see how much of the virus is cleared out over the month (1-100%). The partner of a person with oral or genital herpes needs to be treated as well whether they show signs of it or not, other viruses can allow it into the body's immune system and not show symptoms until stressed.

Herpes-Cold Sore Case Study: A 52 year old female presented with cold sores with stress, hay fever, food allergies, achy joints, heartburn, general anxiety, and high blood pressure. She has a history of pneumonia and bronchitis every year, body tends to be hot, shortness of breath with exertion, sluggish energy, marginal diet-a lot of ice cream and cheese, carbohydrates, 2 cups of coffee and tea daily. She's had a hysterectomy due to endometriosis and fibroids. She had "high end apathy" at work. **Treatment:** Acupuncture, ITM herbs, Homeopathy

Analysis: ITM herb, Women's Treasure was given to perk up her energy and help digest her food. Underlying allergies always stress the body and contribute to a lowered immune system then the herpes breaks out. The acupuncture helped with her anxiety, clear the heat and supported joints which are related to the liver. The liver doesn't like to be hot and all associated organs can react to the heat. (Joints, eyes, finger and toe nails) Candida can make a person damp and makes the environment that viruses like to live in. (See Tangles) She had viruses for bronchitis plus 3 others including mono, chemical toxicity from her work. She worked through the viruses that took 3 months. After 9 treatments her heartburn resolved, no pollen allergies or cold sores from exposure. Her energy was good, anxiety much better, felt calmer, no hot feet or hot flashes, and her blood pressure improved.

Herpes-Genital Case Study: A 35 year old female presented with allergies, chemical pollen and food. She breaks out around menses time and with too much activity. She tends to be hot with cold hands. She likes to sleep a lot and her energy is low. **Treatment:** Acupuncture, ITM herbs, Homeopathy

Analysis: ITM herbs, Restorative and Women's Treasure used to clear heat and charge up the kidneys. Chronic allergies bring the kidneys down that support the immune system. Hormonal changes at menses can make the immune system weak and cause outbreaks. Too much exercise can reduce an already weak kidney that's trying to keep up with making natural steroids and white blood cells for allergy issues. Acupuncture was used to boost the kidneys and immune system. Homeopathy used to take out the herpes virus plus 3 others. Pesticides, Dioxin, Preservative were taken out with Chem X homeopathic; parasites taken out with ITM herb, Omphalia. The viruses need to be taken out in order, easiest 1st usually 1 per month if the immune system is up enough. If the immune system is not then it takes longer so the body can get the nutrition it needs to build the white blood cells to do combat and get the virus and chemicals completely out. (See stages of Illness) It took 3 months and 12 treatments to get the body up so no allergies were manifesting. During that time viruses were taken out 1 at a time plus chemicals and parasites. After 6 months of treatment 1 every 2-3 weeks she had no complaints. After 5 years she had not had a reoccurrence of the genital herpes virus after the initial treatment.

Molluscum Case Study: A 2 year old and 1 year old brother came in with Molluscum. The older brother presented with chronic allergies, ear tubes 2 times, chronic ear infections, exposed to Mono. All vaccines were given. He tends to be warm, red eyes, sleep 10 hours plus a 2-3 hour nap, frustrated and fussy. His diet is balanced and craves sugar. His is on probiotics and 2 medications for allergies.
Treatment: NAET, Homeopathy

Analysis: The older boy had a tremendous amount of allergies for 2 years old. He had inherited or had too many vaccines bringing a weak body down further. We did NAET for 21 items, 1 at a time. During the allergy clearing process he showed up with Molluscum. He had had it for about a month. He got homeopathy 88-88 and it moved out within 2 days. His brother with the same allergy issues started to get it so he was given the homeopathy and it was out within 24 hours. The homeopathy needed to be taken for the duration of the month but the contagious and irritating part was gone within 48 hours. Molluscum looks like Chicken Pox. It is a childhood disease that a child can have for 2 weeks to up to 6 months and is highly contagious. This can be gotten genitally if older. This virus is in the Poxvirdae family. (DNA Pox virus, Monkey Pox) It is spread direct skin to skin contact or object to skin. This virus has been found in 60% of HIV positive people. It mostly affects children between 2-5 years old. It likes the tropics moist and damp heat. The UK saw a 50% rise in contagion from 1998-2008. It cannot be grown in tissue culture or eggs. Antibodies have been found in 80% of patients-1969-1983 saw an increased number of patient visits.

Monkey Pox related to Small Pox related to Molluscum:
2003 Monkey Pox was diagnosed in the United States, a DNA pox virus. There were 47 confirmed cases in Illinois, Indiana, Kansas, Missouri, Ohio and Wisconsin. They came in with animals from Gharia Africa to Texas in 2003; squirrels, rats and doormice contaminated. They infected prairie dogs at an Illinois animal vendors business. Persons purchasing prairie dogs became sick. The CDC (Center of Disease Control U.S government) said they cleaned this up.

Shingles: Treated 20 clients, 15 Females, 5 Males, average age was 53 years old. Shingles usually appear with extreme stress to a person's body, emotions or spirit. It is a virus that radiates down a nerve bundle creating great pain/heat as it makes its way to the nerve ending and erupts into painful blisters. It usually takes a few days to make its way, by this time the client will know if it is shingles. PAIN! This virus starts within the spinal column and pushes out to the end of its length. It is common on back, neck, shoulder and face. It is dangerous on the face because it can radiate into the eye and cause damage. I have had people 4 years later still have pain from the virus. Their immune system couldn't kick it out and it became stuck in a lower layer of the stages of illness. It's precipitated from mono, chronic yeast, surgery /trauma to body, severe allergies, diabetes, cancer, pain from trauma, too much antibiotics, menopause/hot flashes, stress, and depressed immune system. Acupuncture and herbs help immediately with relief. Homeopathy can get it long term so there is no reoccurrence of Shingles.

Shingles Case Study: A 58 year old female presented with shingles, mid back for 2-3 weeks, shingles under right shoulder and radiates from right side of body. She has shortness of breath-she does smoke 1 pack a day. She has insomnia from the pain. Her diet is balanced with 2 cups decaf coffee and 1 soda a day. History of total hysterectomy due to endometriosis. She had mono in high school and almost lost a kidney because of it. She is Allergic to Augmentin and some foods. She is taking Estrogen, Synthroid and Advil for pain. **Treatment:** Acupuncture, ITM herbs

  Analysis: She was put on ITM herb Red peony used especially for Shingles, within the week the lesions were healing. After 1 acupuncture treatment the pain was receding. After 3 treatments there was no more nerve pain, a little possibly at the spots. She managed a very busy accounting business and the stress at tax time precipitated the event.

Strep Virus Case Study: A family of 6 continues to get sick with strep throat repeatedly (Mother, Father, 11, 9, 6 and 8 year olds). The youngest child has Down's syndrome; 2 of the children (11, 6 year olds) are asthmatic. They have taken multiple doses of antibiotics. The 6 year old has taken 5 courses of antibiotics. He is also on Colostrum, probiotics/congeplex, Mangosteen, and has limited dairy. The 6 year old presents symptom free, but turns out to be the carrier.

The family is put on a homeopathy 60-24 for strep virus; 1 month later most of the family is 70-90% clear except for the 6 year old. The family is then put on homeopathy 60-92 for strep virus in the lungs. 2 months later the family is 100% clear of 60-24 and 60-92 strep viruses. The 6 year old still holds on 80% on 60-92. 2 more weeks the 6 year old is clear, no signs or symptoms for any of them the rest of the year.

Strep Virus Case Study: Family of 6 (Father, Mother, 8,6,4,2 years old) constantly gets sick with strep infections. All are checked and the mother is the carrier. She had strep 3 years ago. The 6 and 4 year old have strep a lot; diagnosed 60-24 strep virus in the throat. We used homeopathy. I contacted them 2 months later and everyone was feeling well and no problems.

Chronic Strep Case Study: A 3 year old female presents with mouth sores, bumps on cheeks, thrush, she has a depressed immune system. She had strep throat 3-4 times in a school year. The family is tested and the 6 year old brother is the carrier (host). The rest of the family mother/father, acquire strep virus multiple times per year. All are put on a strep virus homeopathic that clears it out of the family.

She was tested by an M.D for infectous disease and was found negative. She was colicky baby. She had all her childhood vaccines. She tended towards constipation, no temperature, loved fruit and milk products. **Treatment:** NAET, Homeopathy

Analysis: The child's viral load was high at a very young age. Both parents worked at a school, generating of many infections. She presented with allergies from many foods. We corrected the virus issues with homeopathy and the allergy issues with NAET. Eight years later she still reports healthy.

Multiple vaccines may have caused a depressed immune system creating an allergy issue.

Practitioner Treatment Checklist

1- I always check for **Mono and Hepatitis B, C and Herpes**. Mono usually has a presenting fatigue factor. Mono is usually the gateway to other virus loads. It is found in underlying chronic immune problems; Lupus, M.S., Fibromyalgia, Allergies, Cancer, Chemical Sensitivity, some Diabetes type I and II and chronic pain, Autism.

2- Check for **Vaccine load** depending on age, sensitivity to environment, allergies.

Too many vaccines can crash the immune system resulting in allergies to food, chemicals and environmental particles, pollen, mold, autism, ADD. This in turn contributes to arthritis, asthma, autoimmune diseases, fibromyalgia, and chronic pain (sciatic, migraines).

3- Candida is an underlying issue in chronic conditions. Signs and symptoms indicate a diet change needed, for example yeast free, etc. Prolific yeast can contribute to Autoimmune, Fibromyalgia, Allergies, Insomnia, Herpes outbreaks, feeding parasites that contribute to poor gut flora, Diabetes, Diverticulitis, Crohn's, Constipation, Diarrhea, Cancer, Asthma, Arthritis, Skin conditions, chronic sinus infections.

A person can get it from their sexual partner.(Look at Tangles)

4-Anxiety is usually due to a fired up heart but can be back tracked for food allergies, pain, virus load, or the liver is too agitated, hot and has fired up the heart, the emotions and ADD.

5-Parasites are an unseen companion. We deworm our pets, horses and cattle but we never cleanse ourselves. Once a year parasite cleanse is a must. ITM herb Omphalia

6-Cancer cleanse 1 time a year. ITM herb Chihko Curcuma

7-Surgeries can block the meridian lines causing pain and dysfunctions in the body.

8-Cholesterol Medicine can cause leg pain or general Fibromyalgia look at side effects on Medications.

9-Mercury in tailbone can cause leg pain/Fibromyalgia

10-Trace Minerals can fix restless leg syndrome

11-Using Olive Oil on skin can correct dry eyes. Get the cheapest oil with more water in it but high grade and apply after shower on skin. It will absorb and get used elsewhere in the body. Get rid of your lotion; too many chemicals.

12-Prolapse Bladder, Prostate surgery and Urinary Incontinence Treatment same CV1; CV channel threading CV 6,4,3,2.

13-Stress in a person's life always makes any problem they may have worse and can even create the problem. Anxiety, high blood pressure, digestive upsets-acid reflux, Asthma, Constipation, Diarrhea, Fibromyalgia, Headaches, Pain, Insomnia, Menses pain-fibroids, endometriosis, Shingles, Seizures, can be exacerbated or even created to name a few from stress.

14-Check Energy scale (Level of Fatigue) 1 is low and 10 is high. People have a high and a low energy level several times during a day. Check what time applies and check the Chinese Clock to see what meridian has energy in excess or deficient; for example Kidneys are strongest 5-9 pm and weakest 5-9 am. Rest in the am to regenerate the Kidneys. If you have a lot of energy in the evening then you probably have weak kidneys rest them in the morning.

The old monks would wake up at 3 am to meditate because it was the beginning of the day-lung time-and they would do deep breathing exercises so if you have a problem with lungs and wake up at 3 am do some deep breathing. (Refer to chart on website)

15-Clients probably have **Multiple issues.** Treatment starts with most superficial and then works deeper. One can tell the body to go after a certain issue but the body will fix the easiest issue 1st. One problem can segue into another because the body wants to heal the next issue also if available, especially emotional issues trapped in the body. It may appear to the client that you are trying to hold onto them financially, but some people need more work than others so a negotiating scale of treatment should be worked out like every 2-3 weeks or even 1 time a month. If they request a wider spread than that it is not maintainable in my experience.

16-Pulse and Tongue Diagnosis: Chinese Medicine did not rely on blood tests to get the information they needed for evaluation of a client. The practioner looked at the tongue to determine constitutional issues and at the pulses to see transitory but primary conditions. There are a variety of ways to read the pulse and tongue. I was trained TCM-Traditional Chinese Medicine and even some of their trainings are different amongst themselves. I learned a pulse had 9 levels and 22 qualities. Pulse is to be read at the radial pulse at the wrist.

Left Pulse (Yin)	Right Pulse (Yang)
Heart	Lungs
Liver	Digestion
Kidney-Will	Kidney-Will to carry out

When checking a pulse you look for Superficial, Middle/Neutral, and Deep. All should be in neutral position across the fingers. If the point is Superficial it is excess, if it's deep it's deficient. The Kidneys support the other organs in their functioning. There are many books on treatment.

The 22 qualities relate to each point whether they are wiry, soft, weak, hollow, bounding, slippery, deficient; their depth rate, regularity, width, length; smoothness, stiffness and strength. Each of these qualities relate to the functions of the organ. Wiry-superficial in the digestive position indicates heartburn, pain or ulcer. Slippery-superficial in the lungs relates to a cold or sinus/lung congestion. Deep kidney on the left relates to overheating issues; stress, no energy-possible stroke if the liver is too wiry.

The tongue is divided up into sections that represent the specific organs in the body. A practitioner looks at these specific areas and determines if the body is too hot (red), cold (purple) (white), wet, dry, swollen, thin, red tip, red sides, scalloped, plus the coating on the tongue, purple, pale, yellow, spots. All of these attributes mean something at each organ position. Root of the tongue relates to the kidneys. The middle is digestion, the side is Liver/Gall Bladder and the tip is the heart. Lungs are towards the tip.

17-Cupping, Electric Stimulation, Moxa: Moxa is a plant (Artemisia/Mugwort) that is rolled up in a ball and put on an acupuncture needle to bring heat or tonify a specific area of the body. It can come in various forms such as sticks, loose and compressed packs to be applied. It works well when done right. Most practitioners don't have time to utilize it properly on their clients. I send it home with clients as homework.

Electric stimulation acupuncture connects a tens unit to a pair of acupuncture points to open up a meridian area to unblock pain. It can be used anywhere on the body except at the heart. It can also be used to increase mending in an area such as a broken bone, trauma.

Cupping is mostly used on the back or I've seen it on the temples for migraines. It is used to open up deep circulation in the core of the body.

18-NAET (Nambudupad's Allergy Elimination Technique): This was developed by an M.D/Acupuncturist in California who was allergic to everything but rice and broccoli. She fell asleep while eating carrots with needles in and she was better 24 hours later with carrots. She developed a protocol to reprogram the body so it wouldn't be allergic to various irritants. She claims once you clear for something it's cleared for the rest of your life. I had a lot of allergies that got cleared up in 1999 and I haven't had to redo any of them and still feel well.

The process entails holding a vial of the offending allergen, let set 20 minutes/with acupuncture protocol and stay away from allergen for 25 hours. There should be no more allergic reactions with that set of allergens. The time consuming part is that the body can only tolerate so many allergens being processed at a time so it takes a few treatments to resolve the issues depending on how many allergens, problems a person has. It is important that one completes all the NAET treatments of allergy because the few that are not treated become bigger issues to the body later from my experience.

19-Essential Oils can be used on multiple levels of the body from physical pain to mental, emotional, and spiritual issues. These are really not used in Traditional Chinese Medicine and the individual needs to determine what works for them. The smell goes right to the Hypothalamus and can create great change but if one is allergic to the item it can cause allergy issues. Too many at once can cause confusion for the body.

20-Chinese Medicine Balanced Rainbow Diet consists of a Rainbow of foods. One should have a little of every element every day. (See chart on foods for Fire, Earth, Air, Water, and Metal on website) There is also a taste for each Element:

Organ	Sensation	Example	Color
Heart	Bitter	Tea	Red
Liver	Sour	Pickles	Green
Spleen	Sweet	Honey	Yellow
Lungs	Pungent	Salsa	White
Kidney	Salty	Salt	Black

Diet is the hardest to change. I will suggest to a client once, at intake and then with support if they comply; if they don't- I don't continue to push again. It is up to them to change their habits of a lifetime and family lifestyle. It wastes your time and energy to keep pushing and leaves one frustrated.

21-Practitioner Diagnosis is the same for herbs or Acupuncture when working with a client. Some people are so allergic that they can only use Acupuncture. I like to use both to build not only the chemistry/battery but to charge the battery too. (Acupuncture) It takes longer to fix an issue if one is only using Acupuncture.

22-Homeopathy: Homeopathy was developed in the 1800's. The premise is, "like treats like" a tiny dose of the problem will activate the immune system to react and clear the problem out. This theory is what vaccines are based on. Vaccines use larger doses than homeopathy. In the early 1900's Naturopathy/Homeopathy was our primary medicine in the United States. Western Medicine took over and made the previous medicine illegal. Naturopath's are making a comeback and considered Primary Care Physicians in 16 states, as are D.O Doctors of Osteopathy and Chiropractors.

Doctors of Oriental Medicine are considered Primary Care Physicians in New Mexico.

Homeopathy can treat a variety of illnesses. If the medication is not the right one for the issue nothing happens, no harm done. Homeopathy can treat for acute problems and chronic or constitutional problems. If the homeopathy is correct sometimes a healing crisis, **Herxherimer's** can occur, this is the immune system kicking out the offender, it can last for 1-2 days occasionally.

23-Oil Pulling: This is an aryuvedic exercise to clean the mouth and the teeth. A person gets any kind of organic oil; I prefer the taste of coconut. First thing in the morning, 1-2 teaspoons of oil in the mouth and swish for 15-20 minutes. Spit the remains into the trash or a designated jar, one doesn't want to clog the drains long term. This can be done more often in the day. If one is having teeth problems, adding peppermint, cloves or tea tree oil can be added as needed for an issue. This can pull out infection. (See teeth cases) The oil is lipophilic, in other words it attaches to a cell wall of the organisms that make plaque or infections, traps it and then one spits it out. The mouth feels very clean. My hygienist always comments how clean my teeth are and any plaque caught in the mouth is soft. Dental cleaning is not painful anymore.

24-De-stress: Suggestions would be regular yoga, meditation, dancing, riding horses, exercise, gardening, singing, golf, softball, skiing, something fun that relaxes the body. Chinese Medicine believes as active as a person **is** should be balanced with **Quietness**, reading a book, contemplation.

25-Iodine Test: This is a do it yourself Thyroid test to see if one has enough Iodine in their systems to run the thyroid. Coastal people who eat a lot of fish usually get enough iodine but states that don't have those resources have a majority of population with low functioning thyroids that lead to obesity. The test is buying a bottle of iodine at any variety of stores. Apply a swatch of 3 by 3 inches of iodine on your belly or leg in the morning. This swatch should last 24 hours on your skin if it doesn't you need iodine. One can keep putting this iodine on the body as the nutrient that is needed until it lasts 24 hours or you can get iodine pills.

26- PH: An **acidic** body due to diet and habits creates an environment for disease. It can create heat, candida/yeast a hot bed for virus and parasites. An individual should run more alkaline ph 8 or above. PH runs from 1-14. Acidic is less than 7, 7 neutral, 8 and above alkaline. Alcohol, caffeine and sugar are all acid creating. Vegetables and fruit all create an alkaline body. Normal diet should be 80% alkaline and 20% acid. (Check website for list of foods)

Alerts

1-Cholesterol Drugs in Chinese Medicine the eyes are related to the liver. **Statins** can be bad for dry eyes or any eye condition. The statins tend to dry out the liver. They can also cause leg pain, muscle-skeletal pain and restless leg. Statins have been known to go after the myelin sheath, which is made of lipids (fatty acids) that surrounds the nerves of the brain, possibly increasing dementia and Alzheimer's as one ages. It's like stripping an electric wire, then the hot wires cross causing confusion in the brain.

Chemistry dictates if one keeps presenting a chemical to clean a surface and the surface is cleaned, then the product will dig deeper to satisfy the binding of the chemicals together like a lock and key.

Good fats are needed in the system to keep the eyes hydrated, the skin soft, use for breakdown for glucose for energy, keeping the nerves padded so they don't cross wire and cause pain in the body, for example restless leg, neuropathy. Good fat keeps moisture in the skin-keeps one hydrated. Good fat breaks down to make endorphins for the brain, neurotransmitters, natural steroids, such as serotonin, GABA, Acetylcholine, Norepinephrine, Dopamine, (mental health issues) for the body (pain relief)

Good fats include olive oil, butter, avocado oil, coconut oil, no GMO, no chemicals applied. This fat goes to your liver that takes care of the majority of chemical functions in the body. It makes the endorphins and steroids that keep you out of pain and make you think correctly.

2-GMO Foods are fabricated foods. The producers take the original DNA and add snippets of other unrelated DNA into the genetic structure an example would be to add a snippet of a daffodil to rice to help it increase Vitamin D. The thought behind it is to improve the crop so more can be produced to feed the ever expanding population of the world, a noble thought. The problem with this is that our bodies work on a lock and key system where the DNA of the plant has to fit perfectly with the DNA of our bodies for us to absorb it. Chemistry is particular in how chemicals hookup with each other and if the configuration isn't precise our bodies will not accept the plant or animal. So what happens then?

When our bodies are running at less than optimum these products have to be put somewhere in the body. These products get stuck in fatty liver, arthritic joints, plaques on the brain, wherever the system can stash these noxious chemicals away-tumors, breasts, prostate. When the body environment becomes saturated then a disease occurs.

Cleansing is necessary for the body and proper food **Primary**.

Cleansing can be pretty tough to eliminate long standing tissue binding. It takes 7 years to replace all the cells in the body. Candida, parasites, and virus thrive on low frequency cells. These identities are with us to clean up the detritus/garbage in our bodies. If we are not careful they can take over. There has to be a balance of Biotics in the body.

3-Cleansing the body is important to do quite often, major and minor cleansings.

Fruit, Vitamin C is a cleanser. Having fruit in your diet everyday helps the body purge garbage out of it.

Vegetables not only build up the body's store of mineral for Liver functions but also do some cleansing.

Carbohydrates purpose is for energy but can gum up the system.

Protein builds the body and is needed for all DNA functions.

A deep cleanse can be done on every organ on the body (Kroeger "Good Health through Special Diets"). General cleansing helps also. I like to do 1-3 days on watermelon only every 2 hours or as wanted when hungry. This keeps the blood sugar up and clears the lymphatic/cardiac system. One can do purple grape juice in the same manner. Cleansing should be done in the spring or summer.

Warning: When a body is weak too much clearing/cleansing can make a body weaker. It is best to cleanse a little then build more, cleanse a little. Be gentle on yourself it's the only body you have and it's doing the best it can for you! You are the parent take care of this child when it is angry you will surely know it!

4-G.R.A.S. List-(1958 Food Additives Amendment)

(1998) the GRAS list was developed by the F.D.A as a list of ingredients they deemed "Innocuous" neutral to our health when ingesting them. Most of these products have never been tested through the rigorous FDA system. AS of this writing there are 819 pieces on the list. Originally there were 20. The food industry has chemicalized almost every flavor one can imagine and every color. Every year there are big Symposiums around the world to sell these ingredients to Big Business Food Groups. These chemicals are put into base food products to enhance flavor and color. This is cheaper to do than to buy good wholesome "conventional" food that has had fertilizer put back into it to enhance the flavor and color naturally.

Organic food has natural DNA genetics but is expensive seed to grow and not necessarily fertilized because of the overall cost. Conventional seed is fertilized but might not be clear of pesticides/insecticides. The overall cost determines the process; often conventional is better for the body than organic because there is more nutrition in it. I use both.

Fake food is cheaper than wholesome food. Anything in a box is questionable.

Addiction to foods is a hard habit to break, like high fructose corn syrup. Kellogg's did a lot of product testing on cereals in the 50's to make it more enticing, edible to children. They came up with high fructose corn syrup and now it is put in a lot of foods to make it more palatable but also addicting!

If you don't understand the language on the labeling don't eat it! Do your research......

5-Merc Manual Bible of Western Medicine

Merc Manual at least 15 years ago had different normal for cholesterol, blood sugar and blood pressure. Cholesterol at 250 was normal, now the new normal is 200. If the fat content gets too low we have trouble with the brain (increased Alzheimer's and Dementia) pain, eye problems, liver function. (See alert on Cholesterol Drugs!)

Blood sugar normal was 150 and now is 100. At 80 a person can start blacking out/fainting because there isn't enough sugar/energy to keep them awake. The elderly can pass out and fall because of too low sugar.

Blood pressure normal at 20 years of age is 120/80. The press to keep that level throughout life is not appropriate. When adults reach their 40's, 140/90 is appropriate because of wear and tear on the body. The lymphatic, venous, arterial system is well worn and too open at that point for older people. The older body's need more pressure to push the liquids through the body.

All three of these issues if mishandled can cause more problems than the initial issue.

Dietary change can help all of these problems to maintain normality.

Excess medication can cause secondary problems, addictions and dependency that might not have occurred otherwise.

Detoxing and readjusting the body after the fact can take a lot of work and sometimes the bodies can't let go of the addiction. (Fentanyl, opioids, antidepressants, psychotropic)

6-Hot and Cold

The **temperature** of the body can affect the disease process. Our normal temperature is supposed to be at 98.6°. People who run subnormal temperatures long term usually have a thyroid issue or a virus load issue tending toward a chronic immune condition.

In Chinese Medicine when we ask for a feel of someone's temperature we are asking, how do YOU feel? Do you prefer hot or cold drinks? Do you tend to feel hot or cold? How about your hands and feet? We then back engineer into the core condition or constitution of the body.

A lot of conditions/diseases of the body can be created by just being too hot metabolically or too cold. This is a dietary issue.

Heat is created by too much caffeine, sugar, alcohol, smoking, yeast conditions, internal composting of milk products, pharmaceutical drugs.

Cold is created by using too much ice, ice cream, carbohydrates, fake food-additives, preservatives, chemicalized products.

Heat can create and contribute to migraines, endometriosis, fibroids, arthritis, ADD, anxiety, autoimmune, pain, constipation, dry eyes, fibromyalgia, HBP, insomnia, menses issues, seizures, skin problems, UTI's.

Cold can create and contribute to vertigo, thyroid, infertility, osteoporosis, musculoskeletal issues, incontinence, and ears, digestive issues-not cooking the food, diabetes due to poor diet, constipation, asthma, chronic cough, and allergies.

Change the Diet!!!

7-Chinese Herbal Medicine:

Be environmentally conscious when choosing what herbal companies you want to work with. **No endangered species** allowed in remedies!! No **Human** parts! No **Mammal** parts! All of these pieces can be found in renewable resources. We only have **One EARTH** don't misuse the beings on it for your selfish endeavors of self-use.

Biography

Dr. Carol Ragle, Doctor of Oriental Medicine graduated from Oregon College of Oriental Medicine in Portland, Oregon. (1993) She has been practicing Chinese Medicine for 25 years with specialties in Immunology, Allergy, Chronic Immune Deficiencies, Internal Medicine.

She worked in the education fields for 10 years teaching Anatomy/Physiology and Microbiology to nurses and massage therapists with a Bachelor's of Science (1975) in Secondary Education certified Chemistry, Biology, Health Social Studies and Art.

She continued to obtain a Master in Science of Environmental Biology (1980).

She spearheaded 7 Health Fairs in Alaska and was a judge and organizer of Science Fairs winning at National with one of her students.

She worked with 2 different M.D's administering their allergy programs that included testing, vaccines and diet.

She also obtained a RN degree (2010) practicing in a Geriatric Facility and continues to work with Geriatric patients in private practice.

Chinese Medicine has allowed her to work with her clientele in a wide range of disabilities. Chinese Medicine works to integrate an individual not only with themselves but their environment.

She was an Examiner for the State of New Mexico licensing D.O.M.'s for 5 years. She worked with addiction specialists for 15 years. (NADA) She privately practiced 15 years in New Mexico and 10 years in South Dakota. Nationally licensed NCCAOM and NM State licensed.

She has taught meditation for 25 years bringing forth the connectedness of Mind, Body, and Spirit in our own health care. Her focus is learning how to manage your own personal healing for the longevity of your life. Self-care is essential for maintenance of a spirit in a human body. Please take care of it! No one else has more of an investment than you!

Organizations and Books Referenced:

ITM: Institute for Traditional Medicine and Preventative
Health Care
2017 SE. Hawthorne
Portland, Oregon 97214
Website: www.itmonline.org
Phone: (503) 233-4907

Kroeger Herbs Products Inc.
805 Walnut St.
Boulder, Colorado 80302
Website: www.kroegerherb.com
Phone: (800) 516-0690

Standard Process
1200 W. Royal Lee Dr.
Palmyra, Wisconsin 53156
Phone: (800) 848-5061
Website: www.standardprocess.com

Enzymes International Inc.
58 US-51 PO 157
Manitowish Waters, Wisconsin 54545
Website: info@enzymeinternational.com
Phone: (715) 543-8401

NADA: National Acupuncture Detox Association
https://acudetox.com
info@acudetox.com
Phone: (888) 765-NADA (6232)

NEAT: Nambudripad's Allergy Elimination Technique
1440 N. Harbor Blvd. Ste. 105
Fullerton, California 92835
Phone: (714) 523-8900

Organizations and Books Referenced:

The Yeast Connection by: William G. Crook M.D

Eat Right 4 your type: 4 blood types, 4 diets by: Dr. Peter J.D Adams

Dr. Carol Ragle, website: drcarolragledom.shopify.com, email: bluedragondom@gmail.com and Facebook Business Drcarolragledom

Index
Case Study=C.S.

Index
Case Study=C.S.

Index
Case Study=C.S.

Index
Case Study=C.S.

Index
Case Study=C.S.

Index
Case Study=C.S.

Index
Case Study=C.S.

Index
Case Study=C.S.

Index
Case Study=C.S.

Index
Case Study=C.S.

www.ingramcontent.com/pod-product-compliance
Lightning Source LLC
Chambersburg PA
CBHW031811190326
41518CB00006B/291